Multi-Structures and Their Logics, for Computer Science, Computer Engineering, Mathematics, and Philosophy

Matt Insall

2

i

ii

Matt Insall
mattinsall@embarqmail.com

Rolla, Missouri

United States of America

Typeset, using LaTeX, in 12 pt font

ISBN 978-0-6151-5234-9

Printed and bound by lulu.com, inc.

To My Parents, Gene and Bess
To My Wife, Rose Marie
To My Eldest, Thomas Matthew
To My Daughters, Danielle Rose and Mikaela Rose
To My Youngest, Samuel Ruben

iv

Contents

Preface

This book sets forth a theory of "multi-structures", and explores logics and languages that are natural for the study of these mathematical objects. The text is written for upper level undergraduate students and beginning graduate students in Computer Science, Mathematics, and Philosophy, although it is expected that students of other disciplines can benefit from the study of this subject as well. Multi-structures differ from the "structures" of classical logic and model theory in that the arity of a fundamental operation, f_i^{M}, of a multi-structure,

$$\mathbb{M} = \left(M; \left(f_i^{\mathrm{M}} \right)_{i \in I} ; \left(R_j^{\mathrm{M}} \right)_{j \in J} \right)$$

is an ordered pair (m_i, n_i) of nonnegative integers, such that f_i^{M} is a function whose domain is M^{m_i} and whose range is a subset of M^{n_i}. Contrast this with the classical situation, in which m_i is always 1. Thus our multi-structures generalize structures in their "algebraic part". A specific kind of multi-structure is a "multi-algebra", in which the relational type is empty, as with the structures traditionally studied in universal algebra.

In chapter one, we introduce multi-structrues and their morphisms, which we call "homomorphisms", as in traditional universal algebra. We also give some examples of multi-algebras that are not algebras. The focus is on the results from the classical setting that can be generalized to our new setting, and the exercises reflect this bent.

In chapter two, we design languages appropriate for multi-structures, in a manner very similar to the approach of classical logic. A major difference in this chapter from the classical approach is the introduction of projection symbols into the languages, so that the functional symbols with arity (m_i, n_i), where $m_i \neq 1$, can be dealt with effectively. But this poses no real threat, and the rest of the development of the languages is provided in a pedestrian manner. The exercises focus on clarifying the details in the relationships between syntax and semantics, especially in terms of the satisfaction relation.

In chapter three, we present an axiom system for the logic of multi-structures, including the definition of a proof from assumptions. We also describe alphabetic variants in our setting and prove, among other things, a Finiteness Theorem, a Deduction Theorem, a Soundness Theorem, and a Consistency Theorem for the logic of multi-structures.

Chapter four presents the Completeness Theorem for the logic of multi-structures, and the Compactness Theorem. Numerous exercises are provided to demonstrate the applications of these results.

In chapter five, we develop basic notions of a model theory for multi-structures. The notion of a theory of a multi-structure is presented, and the related concept of elementarily equivalent multi-structures is explored. Some Löwenheim-Skolem type results are discussed, and the exercises illustrate the use of notions of elementary equivalence and elementary embeddings. Elementary chains are described, and we prove that the union of an elementary chain of multi-structures is itself an elementary extension of each member of the given chain.

This version of the text is intended for use as a supplementary text

in a beginning logic course, in which the students use also a text
that presents the classical theory, for structures. The author
presents the material (especially chapter one) with an eye toward a
later more extensive development of the theory of multi-algebras and
an associated equational logic for multi-algebras.

The author is grateful to the University of Missouri - Rolla for the
time allowed for the research and composition required in writing
this text. Also, I would like to thank my family, especially my wife,
Rose Marie, and my children, Thomas, Danielle, Mikaela, and
Samuel, for putting up with me while I wrote and typed the
manuscript for this text.

Matt Insall

17 January 2005

Rolla, Missouri

USA

Added for the 2007 edition:

It has recently come to the author's attention that the term

"multi-structure" has a previously established meaning, in

applied mathematics. But astute readers should have no difficulty distinguishing our concept of a "multi-structure" from that which may be encountered elsewhere.

Matt Insall

17 May 2007

Rolla, Missouri

USA

Chapter 1

Fundamentals

1.1   Introduction

In traditional mathematical logic (see [E] and [EFT]),
universal algebra (see [BS]) and model theory, (see [BeS])
the structures one studies are of the form

$$\mathbb{A} = \left(A; \left(c^{\mathbb{A}} \right)_{c \in \mathcal{C}} ; \left(f_i^{\mathbb{A}} \right)_{i \in I} ; \left(R_j^{\mathbb{A}} \right)_{j \in J} \right),$$

where A is a nonempty set, $\mathcal{C}$, I and J are (possibly empty)
sets, and the following hold:

1. for $c \in \mathcal{C}$, $c^{\mathbb{A}} \in A$ is a constant,

2. for $i \in I$, $f_i^{\mathbb{A}} : A^{n_i} \to A$ for some positive integer n_i

   (called the arity of the operation $f_i^{\mathbb{A}}$), and

1

3. for $j \in J$, $R_j^{\mathbb{A}} \subseteq A^{\alpha_j}$, where α_j is a positive integer

(called the arity of $R_j^{\mathbb{A}}$).

In the case that the sets C and J are empty, the structure $\mathbb{A}$ is an algebra, and it is the type of structure studied in universal algebra. Groups, rings and lattices are examples of algebras that are studied in traditional mathematics and its applications. In the case that the sets C and I are empty, the structure $\mathbb{A}$ is a relational structure. An example of a relational structure in computer science and discrete mathematics is a (directed or undirected) graph, and an example of a relational structure in traditional mathematics is a partially ordered set (also called a poset). Consider, however, a vector space V, and an (m, n) tensor $f : V^m \to V^n$. Then a kind of "structure" that is not considered in the traditional mathematical logic setting is the pair

$$(V; f).$$

We will generalize the notion of a structure to include these, as well as many other mathematical objects. We show that many results from traditional logic, model theory, and universal algebra carry over to this general setting, and we provide a logic system for the symbolic study of such new mathematical systems.

1.2 Multi-structures

A multi-structure is a system

$$\mathbb{M} = \left(M; \left(f_i^{\mathrm{M}} \right)_{i \in I}; \left(R_j^{\mathrm{M}} \right)_{j \in J} \right),$$

where M is a nonempty set and

1. for each $i \in I$, there are $m_i, n_i < \omega$ such that

$$f_i^{\mathrm{M}} : M^{m_i} \to M^{n_i},$$

   and

2. for each $j \in J$, there is $\alpha_j < \omega$ such that

$$R_j^{\mathrm{M}} \subseteq M^{\alpha_j}.$$

The mapping $i \mapsto (m_i, n_i)$ from I into $\omega \times \omega$ is the functional type of $\mathbb{M}$, and the mapping $j \mapsto \alpha_j$ from J into ω is the relational type of $\mathbb{M}$.

For each $i \in I$, we say that the function $f_i^{\mathrm{M}} : M^{m_i} \to M^{n_i}$ is an (m_i, n_i)-ary operation on M. (We also say that the pair

(m_i, n_i) is the arity of the operation $f_i^{\mathbb{M}}$. For $j \in J$, we say

that the relation $R_j^{\mathbb{M}} \subseteq M^{\alpha_j}$ is an α_j-ary relation on M, and,

as in traditional model theory, α_j is the arity of $R_j^{\alpha_j}$.

Example 1.2.1 *Let V be a vector space, and let $m, n < \omega$.*

Then an (m, n)-tensor on V (cf [BG]) is a multi-linear

function $f : V^m \to V^n$. Let $M = V$ and $\mathbb{M} = (V; f)$, where f is

a given (m, n)-tensor. Then $\mathbb{M}$ is a multi-structure. Its

functional type is (par abus de notation) (m, n). Its

relational type is empty.

Let $\mathbb{M}$ be a multi-structure with empty relational type.

Then we say that $\mathbb{M}$ is a multi-algebra. The multi-structure

in the preceeding example is an example of a multi-algebra

which is not an algebra.

Example 1.2.2 *Let G be a group with identity i, and define*

operations as follows:

$$f : G^3 \to G^2; (x, y, z) \mapsto (xy, z),$$

$$e : G^0 \to G; e(\emptyset) = i,$$

and

$$\iota : G \to G; \iota(y) = y^{-1}.$$

Let $M = G$ *and let*

$$\mathbb{M} = (M; f, \iota, e).$$

Then $\mathbb{M}$ *is a multi-algebra with the following properties:*

1. $f(f(x, y, z), z) = f(x, f(y, z, z)),$

2. $f(e(\emptyset), y, z) = (y, z),$

3. $f(\iota(y), y, z) = (e(\emptyset), z).$

The (functional) type of $\mathbb{M}$ *is* $((3, 2), (1, 1), (0, 1))$. *We say that* $\mathbb{M}$ *is a* multi-group. *More specifically, we say that any multi-algebra of type* $((3, 2), (1, 1), (0, 1))$ *is a* (left,left)(3,2)-multi-group *if it satisfies the above conditions. One problem of interest is the following:*

Problem 1.2.1 Is there a (left,left) $(3, 2)$*-multi-group* $\mathbb{M}$ *such that for no group* G *is* $\mathbb{M}$ *defined by*

$$f(x, y, z) = (xy, z), e(\emptyset) = i, \iota(y) = y^{-1}?$$

Now we return to the study of general multi-structures. Let $\mathbb{M} = \left(M; \left(f_i^{\mathbb{M}} \right)_{i \in I} ; \left(R_j^{\mathbb{M}} \right)_{j \in J} \right)$ be a multi-structure of functional type $(m_i, n_i)_{i \in I}$ and relational type $(\alpha_j)_{j \in J}$. A subuniverse of $\mathbb{M}$ is a subset N of M such that for any $i \in I$, if $\vec{a} \in N^{m_i}$, then $f_i^{\mathbb{M}}(\vec{a}) \in N^{n_i}$. A substructure of $\mathbb{M}$ is a multi-structure

$$\mathbb{N} = \left(N; \left(f_i^{\mathbb{N}} \right)_{i \in I} ; \left(R_j^{\mathbb{N}} \right)_{j \in J} \right)$$

such that N is a (nonempty) subuniverse of $\mathbb{M}$ and for each $i \in I$ and $j \in J$,

$$f_i^{\mathbb{N}} = f_i^{\mathbb{M}} \cap (N^{m_i} \times N^{n_i})$$

and

$$R_j^{\mathbb{N}} = R_j^{\mathbb{M}} \cap N^{\alpha_j}.$$

In this case, we write $\mathbb{N} \leq \mathbb{M}$. As in classical algebra, universal algebra, and classical model theory, we make use of structure-preserving mappings to study multi-structures and their theories.

Let $\mathbb{M} = \left(M; \left(f_i^{\mathbb{M}} \right)_{i \in I} ; \left(R_j^{\mathbb{M}} \right)_{j \in J} \right)$ and $\mathbb{N} = \left(N; \left(f_i^{\mathbb{N}} \right)_{i \in I} ; \left(R_j^{\mathbb{N}} \right)_{j \in J} \right)$ be multi-structures of functional type $(m_i, n_i)_{i \in I}$ and of relational type $(\alpha_j)_{j \in J}$, and let h be a function from M to N. Then h is a homomorphism from $\mathbb{M}$ into $\mathbb{N}$ if it satisfies the following conditions:

1. for any $i \in I$, if $\vec{a} = (a_1, ..., a_{m_i}) \in M^{m_i}$, and if

$(b_1, ..., b_{n_i}) = f_i^{\mathbb{M}}(\vec{a})$, then

$$(h(b_1), ..., h(b_{n_i})) = f_i^{\mathbb{N}}(h(a_1), ..., h(a_{m_i})),$$

and

2. for any $j \in J$, if $\vec{a} = (a_1, ..., a_{\alpha_j}) \in M^{\alpha_j}$, then $\vec{a} \in R_j^{\mathbb{M}}$ if and only if

$$(h(a_1), ..., h(a_{\alpha_j})) \in R_j^{\mathbb{N}}.$$

If h is a homomorphism from $\mathbb{M}$ into $\mathbb{M}$, then we call h an endomorphism. If h is onto, then we say h is an epimorphism. If it is one-to-one, then it is an embedding, and if it is both one-to-one and onto, then h is an isomorphism. If h is an isomorphism and an endomorphism, then it is an automorphism. If h is a homomorphism from $\mathbb{M}$ onto $\mathbb{N}$, then $\mathbb{N}$ is a homomorphic image of $\mathbb{M}$. As in the case with algebraic structures, intersections of arbitrarily many subuniverses of a multi-structure are again subuniverses, as the following proposition shows.

Proposition 1.2.1 *Let* $\mathbb{M} = \left(M; \left(f_i^{\mathbb{M}} \right)_{i \in I}; \left(R_j^{\mathbb{M}} \right)_{j \in J} \right)$ *be a multi-structure with functional type* $(m_i, n_i)_{i \in I}$, *and let* $\mathfrak{S}$ *be a set of subuniverses of* $\mathbb{M}$. *Then* $\bigcap \mathfrak{S}$ *is a subuniverse of* $\mathbb{M}$.

Proof *Let* $i \in I$ *and let* $\vec{a} \in \left(\bigcap \mathfrak{S} \right)^{m_i}$, *and suppose that* $f_i^{\mathbb{M}}(\vec{a}) \not\subseteq \left(\bigcap \mathfrak{S} \right)^{n_i}$. *Let* $\vec{b} = f_i^{\mathbb{M}}(\vec{a})$, *say* $\vec{b} = (b_1, ..., b_{n_i})$, *and let*

$k \leq n_i$ *be such that* $b_k \notin \bigcap \mathfrak{S}$. *Let* $N \in \mathfrak{S}$ *with* $b_k \notin N$. *Note*

that $\vec{a} \in N^{m_i}$, *so that* $f_i^{\mathbb{M}}(\vec{a}) \in N^{n_i}$. *That is,* $\vec{b} \in N^{n_i}$. *In*

particular, $b_k \in N$, *a contradiction. Thus* $f_i^{\mathbb{M}}(\vec{a}) \in (\bigcap \mathfrak{S})^{n_i}$.

Hence $\bigcap \mathfrak{S}$ *is a subuniverse of* $\mathbb{M}$, *as claimed.*

Given a multi-structure $\mathbb{M}$, we let

$$\mathrm{Sub}(\mathbb{M}) = \{N \subseteq M | N \text{ is a subuniverse of } \mathbb{M}\}.$$

Now, if $X \subseteq M$, then the subuniverse of $\mathbb{M}$ that is generated

by X is given by

$$\mathrm{Sg}(X) = \bigcap \{N \in \mathrm{Sub}(\mathbb{M}) | X \subseteq N\}.$$

Corollary 1.2.1 *Let* $\mathbb{M}$ *be a multi-structure. Then* Sg *is a*

closure operator on M.

Let us digress briefly. Let $\mathcal{P}(M)$ be the power set of M:

$$\mathcal{P}(M) = \{A | A \subseteq M\}.$$

A function $\mathcal{C} : \mathcal{P}(M) \to \mathcal{P}(M)$ is called a closure operator on

M if it satisfies the following conditions:

1. $\mathcal{C}$ is extensive, i.e. if $X \subseteq M$, then $X \subseteq \mathcal{C}(X)$;

2. $\mathcal{C}$ is idempotent, i.e. if $X \subseteq M$, then $\mathcal{C}^2(X) = \mathcal{C}(X)$;

3. $\mathcal{C}$ is monotone, i.e. if $X \subseteq Y \subseteq M$, then $\mathcal{C}(X) \subseteq \mathcal{C}(Y)$.

A subset $X \subseteq M$ such that $\mathcal{C}(X) = X$ is said to be closed with respect to $\mathcal{C}$. It can be shown (see, for example, [BS]) that the set of closed subsets of M is a complete lattice. A closure operator $\mathcal{C}$ on M is said to be an algebraic closure operator if for any $X \subseteq M$,

$$\mathcal{C}(X) = \bigcup \{\mathcal{C}(Y) | Y \subseteq X \text{ and } Y \text{ is finite}\}.$$

We will show that Sg is an algebraic closure operator. Let us first establish some notation. Given a multi-structure

$$\mathbb{M} = \left(M; \left(f_i^{\mathrm{M}} \right)_{i \in I}; \left(R_j^{\mathrm{M}} \right)_{j \in J} \right)$$

of functional type $(m_i, n_i)_{i \in I}$ and relational type $(\alpha_j)_{j \in J}$, if $i \in I$, then

$$\pi_{n_i, k} : M^{n_i} \to M$$

is the projection mapping onto the kth coordinate:

$$\pi_{n_i, k}(b_1, ..., b_{n_i}) = b_k.$$

Given $X \subseteq M$, $i \in I$ and $k \leq n_i$, we let

$$E_{i,k}(X) = X \cup \left\{ \pi_{n_i, k} \left(f_i^{\mathrm{M}}(\vec{a}) \right) | \vec{a} \in X^{m_i} \right\},$$

and then

$$E_i(X) = \bigcup \{E_{i,k}(X) | k \leq n_i\}.$$

Then we let

$$E(X) = \bigcup \{E_i(X) | i \in I\}.$$

Now, set $E^0(X) = X$, and for $n < \omega$, let

$$E^{n+1}(X) = E(E^n(X)).$$

Then set

$$S(X) = \bigcup \{E^n(X) | n < \omega\}.$$

Lemma 1.2.1 *S is extensive.*

Proof *Let $X \subseteq M$. Then $X = E^0(X) \subseteq S(X)$.*

Lemma 1.2.2 *S is idempotent.*

Proof *Let $X \subseteq M$. Then, for $i \in I$ and $k \leq n_i$, we have*

$$p \in E_{i,k}(S(X)) \Rightarrow p \in S(X) \text{ or for some } \vec{a} \in (S(X))^{m_i}, p = \pi_{n_i,k}(f_i^{\mathrm{M}}(\vec{a}))$$

$$\Rightarrow p \in S(X) \text{ or for some } n < \omega \text{ and some } \vec{a} \in (E^n(X))^{m_i}, p = \pi_{n_i,k}(f_i^{\mathrm{M}}(\vec{a}))$$

$$\Rightarrow p \in S(X) \text{ or for some } n < \omega, p \in E_{i,k}(E^n(X))$$

$$\Rightarrow p \in S(X) \text{ or for some } n < \omega, p \in E_i(E^n(X))$$

$$\Rightarrow p \in S(X) \text{ or for some } n < \omega, p \in E(E^n(X))$$

$$\Rightarrow p \in S(X) \text{ or for some } n < \omega, p \in E^{n+1}(X)$$

$$\Rightarrow p \in S(X) \text{ or } p \in S(X) \Rightarrow p \in S(X).$$

Thus $E_{i,k}(S(X)) \subseteq S(X)$, and it follows that $E_i(S(X) \subseteq S(X)$,

and then that $E(S(X)) \subseteq S(X)$. By induction,

$E^n(S(X)) \subseteq S(X)$, for all $n < \omega$, so we have $S(S(X)) \subseteq S(X)$.

But S is extensive, so $S(X) \subseteq S(S(X))$. Thus $S^2(X) = S(X)$.

Hence S is idempotent, as claimed.

Lemma 1.2.3 *S is monotone.*

Proof *Let $X \subseteq Y \subseteq M$. If $i \in I$ and $k \leq n_i$ and $p \in E_{i,k}(X)$,*

then either $p \in X$, in which case $p \in Y$, or for some $\vec{a} \in X^{m_i}$

and $k \leq n_i$, we have $\pi_{n_i,k}(f_i^{\mathrm{M}}(\vec{a})) = p$. In the latter case,

$\vec{a} \in Y^{m_i}$, so it follows that $p \in E_{i,k}(Y)$. Thus we have shown

that

$$E_{i,k}(X) \subseteq E_{i,k}(Y).$$

It follows that $E_i(X) \subseteq E_i(Y)$, and then that $E(X) \subseteq E(Y)$.

An induction argument shows that if $n < \omega$, then

$E^n(X) \subseteq E^n(Y)$. Thus it follows that

$$S(X) \subseteq S(Y),$$

so S is monotone, as claimed.

Corollary 1.2.2 S *is a closure operator on* M.

Theorem 1.2.1 S *is an algebraic closure operator on* M.

Proof *Let $X \subseteq M$ and $p \in S(X)$, so that for some $n < \omega$,*
$p \in E^n(X)$. We will find a finite set $Y \subseteq X$ such that
$p \in S(Y)$. If $p \in X$, then $p \in E^0(\{p\})$, so that $p \in S(\{p\})$,
whence in this case, we may take $Y = \{p\}$. Thus, suppose
that $p \notin X$, and that $k < n$ implies that if $q \in E^k(X)$, then
for some finite set $Y \subseteq X$, $q \in S(Y)$. Since $p \in E^n(X)$, we
have $p \in E(E^{n-1}(X))$, so let $i \in I$ with $p \in E_i(E^{n-1}(X))$. Let
$k \leq n_i$ with $p \in E_{i,k}(E^{n-1}(X))$, and let $\vec{a} \in (E^{n-1}(X))^{m_i}$ with
$p = \pi_{n_i,k}(f_i^{\mathrm{M}}(\vec{a}))$. Let a_1, ..., a_{m_i} be the components of $\vec{a}$, and
let $Y_1, ..., Y_{m_i} \subseteq X$ be finite with $a_1 \in S(Y_1)$, ..., $a_{m_i} \in S(Y_{m_i})$.
Let $Y = \bigcup_{l=1}^{m_i} Y_l$. Then $a_1, ..., a_{m_i} \in S(Y)$, so let $\nu < \omega$ with
$a_1, ..., a_{m_i} \in E^{\nu}(Y)$. Then $p \in E_{i,k}(E^{\nu}(Y))$, so that
$p \in E_i(E^{\nu}(Y))$, whence $p \in E(E^{\nu}(Y))$. Thus $p \in E^{\nu+1}(Y)$, so
$p \in S(Y)$. By induction, we have that if $p \in E^n(X)$, then for
some finite $Y \subseteq X$, $p \in S(Y)$. It follows that if $p \in S(X)$,
then for some finite $Y \subseteq X$, $p \in S(Y)$. Thus

$$S(X) \subseteq \bigcup \{S(Y) | Y \subseteq X \text{ and } Y \text{ is finite}\}.$$

*Since S is monotone, we have $S(Y) \subseteq S(X)$ **for every***
*(finite) $Y \subseteq X$, **so that***

$$\bigcup \{S(Y) | Y \subseteq X \text{ and } Y \text{ is finite}\} \subseteq S(X).$$

Thus

$$S(X) = \bigcup \{S(Y) | Y \subseteq X \text{ and } Y \text{ is finite}\}.$$

*It follows that S **is an algebraic closure operator on** M.*

The following lemma is easy to prove, so we leave its proof
to the reader. We use it in proving the subsequent theorem.

Lemma 1.2.4 *If $k_1, k_2 < \omega$, then there is $k < \omega$ such that*
*$E^{k_1}(X) \cup E^{k_2}(X) \subseteq E^k(X)$. **It follows that if** $k_1, ..., k_m < \omega$,*
then there is $k < \omega$ such that $E^{k_1}(X) \cup ... \cup E^{k_m}(X) \subseteq E^k(X)$.

Theorem 1.2.2 *Let $X \subseteq M$. **Then** $S(X)$ **is a subuniverse of***
M.

Proof *Let $i \in I$ and $\vec{a} \in (S(X))^{m_i}$, **say***

$$\vec{a} = (a_1, ..., a_{m_i}).$$

Let $k_1, ..., k_{m_i} < \omega$ **with**

$$a_1 \in E^{k_1}(X), ..., a_{m_i} \in E^{k_{m_i}}(X),$$

and let $k < \omega$ with

$$E^{k_1}(X) \cup ... \cup E^{k_{m_i}}(X) \subseteq E^k(X).$$

Then $\vec{a} \in (E^k(X))^{m_i}$. Let $\vec{b} = (b_1, ..., b_{n_i})$ be $f_i^{\mathbb{M}}(\vec{a})$. Then for $l \leq n_i$, **we have**

$$b_l = \pi_{n_i, l}(f_i^{\mathbb{M}}(\vec{a})) \in E_{i,l}(E^k(X)).$$

Thus $b_1, ..., b_{n_i} \in E_i(E^k(X))$, so that

$$\vec{b} \in (E(E^k(X))^{n_i} = (E^{k+1}(X))^{n_i}.$$

Hence $\vec{b} \in (S(X))^{n_i}$, i.e.

$$f_i^{\mathbb{M}}(\vec{a}) \in (S(X))^{n_i}.$$

It follows that $S(X)$ is a subuniverse of $\mathbb{M}$, as claimed.

Corollary 1.2.3 *If $X \subseteq M$, then $\mathrm{Sg}(X) \subseteq S(X)$.*

Proof *We have*

$$S(X) \in \mathrm{Sub}(\mathbb{M})$$

and $X \subseteq S(X)$, so that

$$\bigcap \{N \subseteq M | N \text{ is a subuniverse of } \mathbb{M} \text{ and } X \subseteq N\} \subseteq S(X).$$

But

$\mathrm{Sg}(X) = \bigcap \{N \subseteq M | N \text{ is a subuniverse of } \mathbb{M} \text{ and } X \subseteq N\}$, *so we are done.*

Theorem 1.2.3 *Let $N \in \mathrm{Sub}(\mathbb{M})$ with $X \subseteq N$. Then*
$S(X) \subseteq N$. Consequently, $\mathrm{Sg}(X) = S(X)$, so Sg is an
algebraic closure operator.

Proof *Let $i \in I$ and $k \leq n_i$, and let $e \in E_{i,k}(X)$. If $e \in X$,*
then $e \in N$, so suppose that

$$e = \pi_{n_i,k}\big(f_i^{\mathbb{M}}(\vec{a})\big),$$

where $\vec{a} \in X^{m_i}$. Let $\vec{b} = (b_1, ..., b_{n_i})$ be $f_i^{\mathbb{M}}(\vec{a})$, so that $e = b_k$.
Then $\vec{a} \in N^{m_i}$, so $\vec{b} \in N^{n_i}$. It follows that $e = b_k$ is in N.
Thus $E_{i,k}(X) \subseteq N$. Consequently,

$$E_i(X) \subseteq N, \text{ and so } E(X) \subseteq N.$$

An induction argument yields $E^n(X) \subseteq N$ for all $n < \omega$.
Thus $S(X) \subseteq N$, as desired. It follows that $S(X) \subseteq \mathrm{Sg}(X)$.
Hence $S(X) = \mathrm{Sg}(X)$. Since S is an algebraic closure
operator, the desired conclusion holds.

We will pause now for some exercises.

1.3   Congruences

In classical modern algebra, groups, rings, vector spaces,
modules, and lattices possess "nice" substructures that are

the "kernels" of homomorphisms. For groups, normal

subgroups suffice. For rings, the pertinent subrings are

ideals. In a vector space, every subspace is a "kernel" of a

homomorphism (i.e. of a linear transformation). A similar

situation holds for modules. In lattice theory, one has an

option: either one uses ideals as kernels or one uses filters

as kernels. But in general algebra and model theory, the

appropriate object to call a "kernel" of a homomorphism is

a congruence, which we shall define shortly. A congruence

in universal algebra is a subalgebra of $\mathbb{A}^2$ for some given

algebra $\mathbb{A}$, which is also an equivalence relation. But in

model theory, one must also deal with the relations that

correspond to the predicate symbols of the given language,

and in the theory of multi-structures, we must also deal

with the fact that our operations do not necessarily map

into M.

Let $\mathbb{M} = \left(M; \left(f_i^{\mathrm{M}} \right)_{i \in I}; \left(R_j^{\mathrm{M}} \right)_{j \in J} \right)$ be a multi-structure, and let

$\theta \subseteq M^2$ be a binary relation on the set M. Then θ is a

congruence on $\mathbb{M}$ provided that it is an equivalence relation

such that

    1. for each $i \in I$, if $\vec{a} = (a_1, ..., a_{m_i}), \vec{a}' = (a'_1, ..., a'_{m_i}) \in M^{m_i}$ are

such that $(a_1, a_1') \in \theta$, ..., $(a_{m_i}, a_{m_i}') \in \theta$, and if

$f_i^M(\vec{a}) = \vec{b} = (b_1, ..., b_{n_i})$ and $f_i^M(\vec{a}') = \vec{b}' = (b_1', ..., b_{n_i}')$, then

$(b_1, b_1') \in \theta$, ..., $(b_{n_i}, b_{n_i}') \in \theta$, and

2. for each $j \in J$, if $\vec{a} = (a_1, ..., a_{\alpha_j}) \in R_j^M$ and if

$\vec{a}' = (a_1', ..., a_{\alpha_j}') \in M^{\alpha_j}$ satisfies $(a_1, a_1') \in \theta$, ...,

$(a_{\alpha_j}, a_{\alpha_j}') \in \theta$, then $\vec{a}' \in R_j^M$.

Proposition 1.3.1 *Let* M *be a multi-structure, and let* θ *be a congruence on* M. *Then* θ *is a substructure of* $M^2 = M \times M$. *Conversely, if* θ *is an equivalence relation that is a substructure of* M^2, *and if also* θ *satisfies*

- *for each* $j \in J$, *if* $\vec{a} = (a_1, ..., a_{\alpha_j}) \in R_j^M$ *and if*

$\vec{a}' = (a_1', ..., a_{\alpha_j}') \in M^{\alpha_j}$ *satisfies* $(a_1, a_1') \in \theta$, ...,

$(a_{\alpha_j}, a_{\alpha_j}') \in \theta$, *then* $\vec{a}' \in R_j^M$,

then θ *is a congruence on* M.

Proof *We show that any congruence on* M *is a substructure of* M^2, *and leave the converse to the reader. Thus, let* $i \in I$ *and let* $\vec{a} = (a_1, ..., a_{m_i}) \in \theta^{m_i}$, *say*

$\vec{a} = ((a_1^{(1)}, a_1^{(2)}), ..., (a_{m_i}^{(1)}, a_{m_i}^{(2)}))$, *where* $(a_1^{(1)}, a_1^{(2)})) \in \theta$, ...,

$(a_{m_i}^{(1)}, a_{m_i}^{(2)}) \in \theta$. *Let* $\vec{a}^{(1)} = (a_1^{(1)}, ..., a_{m_i}^{(1)})$ *and* $\vec{a}^{(2)} = (a_1^{(2)}, ..., a_{m_i}^{(2)})$, *and let $\vec{b}^{(1)} = f_i^{\mathrm{M}}(\vec{a}^{(1)})$ and $\vec{b}^{(2)} = f_i^{\mathrm{M}}(\vec{a}^{(2)})$, say $\vec{b}^{(1)} = (b_1^{(1)}, ..., b_{n_i}^{(1)})$ and $\vec{b}^{(2)} = (b_1^{(2)}, ..., b_{n_i}^{(2)})$. Since θ is a congruence on $\mathbb{M}$, it follows that $(b_1^{(1)}, b_1^{(2)}), ..., (b_{n_i}^{(1)}, b_{n_i}^{(2)}) \in \theta$. But this means that*

$$f_i^{\mathrm{M} \times \mathrm{M}}(\vec{a}) = ((b_1^{(1)}, b_1^{(2)}), ..., (b_{n_i}^{(1)}, b_{n_i}^{(2)})) \in \theta^{n_i}.$$

It follows then that θ is a substructure of $\mathbb{M} \times \mathbb{M}$, as claimed.

1.4   Further Reading

Readers who like the algebraic flavour of our description of multi-structures may enjoy reading also [BeS], [BS], [H], and other model theory or universal algebra texts. We note that the theory of multi-algebras can be developed further, along the same lines as [BS].

A more traditional logic approach can be found in [M], and an approach relating more to the computer science study of languages is available in [EFT]. Our treatment of languages begins in chapter 2 of the present work, and if a reader feels the need to prepare for our chapter 2, [M] and [EFT] and [E] (among others) are recommended.

The mathematical inspiration for the definition of a

multi-structure comes from the theory of tensors, and a plethora of tensor analysis texts abounds. To get a feel for the notion of a tensor and its connection to multi-structures, readers may want to visit any of several texts on tensors and tensor analysis, such as [BG].

1.5 Future

Future editions will expand the sections in various ways. First, more examples will be given, and developed. Second, examples of multi-structures from Computer Science, Computer Engineering, and Philosophy will be added. Third, the study of congruences on multi-structures, begun in section 3, will be expanded. Their connection to homomorphisms will be elaborated, as is done in sundry universal algebra tomes. Other sections will be added, elucidating the concept of a multi-algebra, and applications in various settings.

1.6   Errors; Suggestions

Only either the author or the publisher is to be blamed for any errata the esteemed reader may find. Please tell them (at mattinsall@embarqmail.com), so that future editions can incorporate such corrections. (In particular, neither lulu.com, nor its employees or subsidiaries, etc, shall be held responsible for any mathematical or scientific errors in this text.

Similarly, please send any suggestions you may have for expansion and improvement of this text. All such suggestions will be considered seriously, and will be gratefully received.

1.7   Acknowledgements

Many people in my personal and academic life helped make this project advance to publication. In some cases, contributions are more direct than in others. I will not delineate all of these cases, however, I will mention some. First, in my personal life, the patience and consideration of my family and friends has been a blessing. Second, colleagues have listened patiently when I explain to them some aspect of logic or model theory. In particular, Jeffrey Dalton listened to some early presentations I gave him about multi-algebras. His comments were invaluable.

Third, many years ago, I studied logic from Klaus Kaiser. His admonitions still linger in my mind, and help to guide my construction of this work.

Of course, I want to thank the people @ lulu.com for providing print on demand services, which help to keep this book affordable.

1.8   Exercises

Exercise 1.8.1 *Let* $\mathbb{M} = \left(M; \left(f_i^{\mathrm{M}} \right)_{i \in I} ; \left(R_j^{\mathrm{M}} \right)_{j \in J} \right)$ *be a multi-structure, let* N *be a set, and let* $h : M \to N$ *be a one-to-one and onto function. Show that there is a unique multi-structure*

$$\mathbb{N} = \left(N; \left(f_i^{\mathrm{N}} \right)_{i \in I} ; \left(R_j^{\mathrm{N}} \right)_{j \in J} \right)$$

such that h *is an isomorphism from* $\mathbb{M}$ *onto* $\mathbb{N}$.

Exercise 1.8.2 *Let* $\mathbb{M}$ *and* $\mathbb{N}$ *be multi-structures with an embedding* $h : M \to N$ *from* $\mathbb{M}$ *into* $\mathbb{N}$. *Show that there is a multi-structure* $\mathbb{K}$ *that is isomorphic to* $\mathbb{N}$ *such that* $\mathbb{M} \leq \mathbb{K}$. *(Multi-structures* $\mathbb{K}$ *and* $\mathbb{N}$ *are* isomorphic *if there is an isomorphism from* $\mathbb{K}$ *onto* $\mathbb{N}$.*)*

Exercise 1.8.3 *Let* $\mathbb{K}$, $\mathbb{M}$, *and* $\mathbb{N}$ *be multi-structures of the same type, and let* $h : \mathbb{K} \to \mathbb{M}$ *and* $g : \mathbb{M} \to \mathbb{N}$ *be homomorphisms. Show that* $g \circ h$ *is a homomorphism from* $\mathbb{K}$ *into* $\mathbb{N}$.

Exercise 1.8.4 *Let* $\mathbb{M}$ *and* $\mathbb{N}$ *be multi-structures of the same type, and let* $h : \mathbb{M} \to \mathbb{N}$ *be a homomorphism. Show* *that the set*

$$h[M] = \{h(a)|a \in M\}$$

is a subuniverse of $\mathbb{N}$.

Exercise 1.8.5 *Let* $\mathbb{M}$ *and* $\mathbb{N}$ *be multi-structures of the same type, and let* $h : \mathbb{M} \to \mathbb{N}$ *be a homomorphism. Let* $\mathbb{K} \leq \mathbb{N}$. *Show that the set*

$$h^{-1}[K] = \{a \in M|h(a) \in K\}$$

is a subuniverse of $\mathbb{M}$.

Exercise 1.8.6 *Let* $\mathbb{M}$ *and* $\mathbb{N}$ *be multi-structures of the same type, let* $h : \mathbb{M} \to \mathbb{N}$ *be a homomorphism, and let* $\mathbb{K} \leq \mathbb{M}$. *Show that the set*

$$h[K] = \{h(a)|a \in K\}$$

is a subuniverse of $\mathbb{N}$.

Exercise 1.8.7 *Let* $\mathbb{M}$ *and* $\mathbb{N}$ *be multi-structures of the same type, let* $h : \mathbb{M} \to \mathbb{N}$ *be a homomorphism, and let* $X \subseteq M$. *Show that*

$$h[\mathrm{Sg}(X)] = \mathrm{Sg}(h[X]).$$

Exercise 1.8.8 *Let* $\mathbb{M}$ *and* $\mathbb{N}$ *be multi-structures of the same type, and let* $h : \mathbb{M} \to \mathbb{N}$ *be a homomorphism. Define*

$$\theta = \{(a, b) \in M^2 | h(a) = h(b)\}.$$

Show that θ *is* equivalence relation. *(That is, show that (i)* θ *is* reflexive, *meaning that* $(a, a) \in \theta$ *for* $a \in M$, *(ii)* θ *is* symmetric, *meaning that* $(a, b) \in \theta \Rightarrow (b, a) \in \theta$, *and* θ *is* transitive, *meaning that if* $(a, b), (b, c) \in \theta$, *then* $(a, c) \in \theta$.*)*

Exercise 1.8.9 *Let* $\mathfrak{S}$ *be a collection of multi-structures, and denote by* $\cong$ *the isomorphism relation on* $\mathfrak{S}$:

$$\mathbb{M} \cong \mathbb{N} \text{ if and only if } \mathbb{M} \text{ is isomorphic to } \mathbb{N}.$$

Show that $\cong$ *is an equivalence relation on* $\mathfrak{S}$.

Exercise 1.8.10 *Recall that if* $\mathbb{M}$ *and* $\mathbb{N}$ *are multi-structures, then we write* $\mathbb{M} \leq \mathbb{N}$ *iff* $\mathbb{M}$ *is a substructure of* $\mathbb{N}$. *Let* $\mathfrak{S}$ *be a collection of multi-structures, and show that the relation* $\leq$ *is a partial ordering on* $\mathfrak{S}$. *(Note: A relation* R *is a* partial ordering *on a set* X *if it is (i)* reflexive, *(ii)* anti-symmetric, *meaning that if* aRb *and* bRa, *then* $a = b$, *and (iii)* transitive.*)*

Exercise 1.8.11 *Find multi-structures* $\mathbb{M}$ *and* $\mathbb{N}$ *such that each is a homomorphic image of the other, but* $\mathbb{M}$ *and* $\mathbb{N}$ *are not isomorphic.*

Exercise 1.8.12 *Let* $\mathbb{M} = \left(M; \left(f_i^{\mathbb{M}} \right)_{i \in I} ; \left(R_j^{\mathbb{M}} \right)_{j \in J} \right)$ *and* $\mathbb{N} = \left(N; \left(f_i^{\mathbb{N}} \right)_{i \in I} ; \left(R_j^{\mathbb{N}} \right)_{j \in J} \right)$ *be multi-structures of the same type. Define*

$$\mathbb{M} \times \mathbb{N} = \left(M \times N; \left(f_i^{\mathbb{M} \times \mathbb{N}} \right)_{i \in I} ; \left(R_j^{\mathbb{M} \times \mathbb{N}} \right)_{j \in J} \right)$$

by the following:

1. *Given* $i \in I$ *and* $\vec{a} \in (M \times N)^{m_i}$, *say*

$$\vec{a} = \left(\left(a_1^{(1)}, a_1^{(2)} \right), ..., \left(a_{m_i}^{(1)}, a_{m_i}^{(2)} \right) \right),$$

   if $\vec{b}^{(1)} = f_i^{\mathbb{M}} \left(a_1^{(1)}, ..., a_{m_i}^{(1)} \right)$, *and* $\vec{b}^{(2)} = f_i^{\mathbb{N}} \left(a_1^{(2)}, ..., a_{m_i}^{(2)} \right)$, *say*

$$\vec{b}^{(1)} = \left(b_1^{(1)}, ..., b_{n_i}^{(1)} \right) \text{ and } \vec{b}^{(2)} = \left(b_1^{(2)}, ..., b_{n_i}^{(2)} \right),$$

   then

$$f_i^{\mathbb{M} \times \mathbb{N}}(\vec{a}) = \left(\left(b_1^{(1)}, b_1^{(2)} \right), ..., \left(b_{n_i}^{(1)}, b_{n_i}^{(2)} \right) \right).$$

2. *Given* $j \in J$, *define* $R_j^{\mathbb{M} \times \mathbb{N}}$ *via the following rule*

$$\left(\left(a_1^{(1)}, a_1^{(2)} \right), ..., \left(a_{\alpha_j}^{(1)}, a_{\alpha_j}^{(2)} \right) \right) \in R_j^{\mathbb{M} \times \mathbb{N}}$$

if and only if

$$\left(a_1^{(1)}, ..., a_{\alpha_j}^{(1)}\right) \in R_j^{\mathbb{M}}$$

and

$$\left(a_1^{(2)}, ..., a_{\alpha_j}^{(2)}\right) \in R_j^{\mathbb{N}}.$$

Prove that $\mathbb{M} \times \mathbb{N}$ *is a multi-structure of the same type as* $\mathbb{M}$ *and* $\mathbb{N}$.

Exercise 1.8.13 *Let* $\mathbb{M}$ *and* $\mathbb{N}$ *be multi-structures of the same type. Is* $\mathbb{M} \times \mathbb{N}$ *equal to* $\mathbb{N} \times \mathbb{M}$?

Exercise 1.8.14 *Let* $\mathbb{K}$, $\mathbb{M}$, *and* $\mathbb{N}$ *be multi-structures of the same type. Which of the following are equal:* $(\mathbb{K} \times \mathbb{M}) \times \mathbb{N}$, $\mathbb{K} \times (\mathbb{M} \times \mathbb{N})$, $\mathbb{M} \times (\mathbb{K} \times \mathbb{N})$, $(\mathbb{N} \times \mathbb{M}) \times \mathbb{K}$?

Exercise 1.8.15 *Let* $\mathbb{M}$ *and* $\mathbb{N}$ *be multi-structures of the same type. Is* $\mathbb{M} \times \mathbb{N}$ *isomorphic to* $\mathbb{N} \times \mathbb{M}$?

Exercise 1.8.16 *Let* $\mathbb{K}$, $\mathbb{M}$, *and* $\mathbb{N}$ *be multi-structures of the same type. Which of the following are isomorphic:* $(\mathbb{K} \times \mathbb{M}) \times \mathbb{N}$, $\mathbb{K} \times (\mathbb{M} \times \mathbb{N})$, $\mathbb{M} \times (\mathbb{K} \times \mathbb{N})$, $(\mathbb{N} \times \mathbb{M}) \times \mathbb{K}$?

Exercise 1.8.17 *Let* $\mathbb{K}$, $\mathbb{M}$, *and* $\mathbb{N}$ *be multi-structures of the same type, say*

$$\mathbb{K} = \left(K; \left(f_i^{\mathbb{K}}\right)_{i \in I}; \left(R_j^{\mathbb{K}}\right)_{j \in J}\right),$$

$$\mathbb{M} = \left(M; \left(f_i^{\mathrm{M}} \right)_{i \in I}; \left(R_j^{\mathrm{M}} \right)_{j \in J} \right),$$

and

$$\mathbb{N} = \left(N; \left(f_i^{\mathrm{N}} \right)_{i \in I}; \left(R_j^{\mathrm{N}} \right)_{j \in J} \right).$$

Define

$$\mathbb{K} \times \mathbb{M} \times \mathbb{N} = \left(K \times M \times N; \left(f_i^{\mathbb{K} \times \mathbb{M} \times \mathbb{N}} \right)_{i \in I}; \left(R_j^{\mathbb{K} \times \mathbb{M} \times \mathbb{N}} \right)_{j \in J} \right)$$

as follows:

1. **Given** $\vec{a} = \left(\left(a_1^{(1)}, a_1^{(2)}, a_1^{(3)} \right), ..., \left(a_{m_i}^{(1)}, a_{m_i}^{(2)}, a_{m_i}^{(3)} \right) \right)$ *in*

   $(K \times M \times N)^{m_i}$, *if*

   $$\vec{b}^{(1)} = \left(b_1^{(1)}, ..., b_{n_i}^{(1)} \right) = f_i^{\mathbb{K}} \left(a_1^{(1)}, ..., a_{m_i}^{(1)} \right),$$

   $$\vec{b}^{(2)} = \left(b_1^{(2)}, ..., b_{n_i}^{(2)} \right) = f_i^{\mathrm{M}} \left(a_1^{(2)}, ..., a_{m_i}^{(2)} \right),$$

   and

   $$\vec{b}^{(3)} = \left(b_1^{(3)}, ..., b_{n_i}^{(3)} \right) = f_i^{\mathrm{N}} \left(a_1^{(3)}, ..., a_{m_i}^{(3)} \right),$$

   then let

   $$f_i^{\mathbb{K} \times \mathbb{M} \times \mathbb{N}}(\vec{a}) = \left(\left(b_1^{(1)}, b_1^{(2)}, b_1^{(3)} \right), ..., \left(b_{n_i}^{(1)}, b_{n_i}^{(2)}, b_{n_i}^{(3)} \right) \right).$$

2. **Given** $j \in J$, **let** $R_j^{\mathbb{K} \times \mathbb{M} \times \mathbb{N}}$ **be defined via the following**

   rule

   $$\left(\left(a_1^{(1)}, a_1^{(2)}, a_1^{(3)} \right), ..., \left(a_{\alpha_j}^{(1)}, a_{\alpha_j}^{(2)}, a_{\alpha_j}^{(3)} \right) \right) \in R_j^{\mathbb{K} \times \mathbb{M} \times \mathbb{N}}$$

if and only if

$$\left(a_1^{(1)}, ..., a_{\alpha_j}^{(1)}\right) \in R_j^{\mathbb{K}},$$

$$\left(a_1^{(2)}, ..., a_{\alpha_j}^{(2)}\right) \in R_j^{\mathbb{M}},$$

and

$$\left(a_1^{(3)}, ..., a_{\alpha_j}^{(3)}\right) \in R_j^{\mathbb{N}}.$$

Prove that $\mathbb{K} \times \mathbb{M} \times \mathbb{N}$ *is a multi-structure of the same type as* $\mathbb{K}$, $\mathbb{M}$, *and* $\mathbb{N}$.

Exercise 1.8.18 *Let* $\mathbb{K}$, $\mathbb{M}$, *and* $\mathbb{N}$ *be multi-structures of the same type. Does* $(\mathbb{K} \times \mathbb{M}) \times \mathbb{N}$ *equal* $\mathbb{K} \times \mathbb{M} \times \mathbb{N}$?

Exercise 1.8.19 *Let* $\mathbb{K}$, $\mathbb{M}$, *and* $\mathbb{N}$ *be multi-structures of the same type. Is* $(\mathbb{K} \times \mathbb{M}) \times \mathbb{N}$ *isomorphic to* $\mathbb{K} \times \mathbb{M} \times \mathbb{N}$?

Exercise 1.8.20 *Let* R *denote the set of real numbers, and let* C *denote the set of complex numbers. Define* $f : R^2 \rightarrow R^2$ *and* $g : C^2 \rightarrow C^2$ *by*

$$f(x_1, x_2) = (x_1 + 2x_2, x_2) \text{ and } g(z_1, z_2) = (z_1 + 2z_2, z_2),$$

and let $\mathbb{M} = (R; f)$ *and* $\mathbb{N} = (C; g)$. *Find all homomorphisms* h *from* $\mathbb{M}$ *into* $\mathbb{N}$.

Exercise 1.8.21 *Let* $f, g : [0, 1]^3 \to [0, 1]^2$ *be defined by*

$$f(x_1, x_2, x_3) = \left(x_1 x_2, \frac{1}{2} x_3\right) \text{ and } g(x_1, x_2, x_3) = \left(\frac{1}{3} x_1, x_2 x_3^2\right),$$

and then let

$$\mathbb{M} = ([0, 1]; f) \text{ and } \mathbb{N} = ([0, 1]; g).$$

Find all homomorphisms h from $\mathbb{M}$ into $\mathbb{N}$.

Exercise 1.8.22 *Let R denote the set of real numbers, and let C denote the set of complex numbers. Define $f : R^2 \to R^2$ and $g : C^2 \to C^2$ by*

$$f(x_1, x_2) = (x_1 + 2x_2, x_2) \text{ and } g(z_1, z_2) = (z_1 + 2z_2, z_2).$$

Let $\mathbb{M} = (R^2; f)$ and $\mathbb{N} = (C^2; g)$. Find all $a, b, c, d \in C$ such that if $h : R^2 \to C^2$ is defined by

$$h(x_1, x_2) = (ax_1 + bx_2, cx_1 + dx_2),$$

then h is a homomorphism from $\mathbb{M}$ into $\mathbb{N}$. Can you find a homomorphism from $\mathbb{M}$ into $\mathbb{N}$ that is not of this form?

Exercise 1.8.23 *Let $+$ denote the usual addition operation on the set R of real numbers, and let $f : R^4 \to R^2$ be given by*

$$f(a_1, a_2, a_3, a_4) = (a_1 + a_3, a_2 + a_4).$$

Let $\mathbb{M} = (R; +)$ *and let* $\mathbb{N} = (R; +, f)$. *(Note: In this case,* $\mathbb{M}$ *is said to be a* reduct *of* $\mathbb{N}$.*) Show that if* $h : R \to R$, *then* h *is an endomorphism of* $\mathbb{M}$ *iff* h *is an endomorphism of* $\mathbb{N}$.

Exercise 1.8.24 *Let* $\mathbb{M} = (M; f, \iota, e)$ *be a (left,left)(3,2)-multi-group. Show that if* $y, z \in M$, *then*

$$f(\iota(y), f(y, z, z)) = (z, z).$$

Exercise 1.8.25 *Let* C *denote the set of complex numbers, let* $T : C^{(2 \times 1)} \times C^{(2 \times 1)} \to C^{(2 \times 1)} \times C^{(2 \times 1)}$ *be the* bilinear $(2, 2)$ *tensor on the complex vector space* $C^{(2 \times 1)}$ *that is defined by*

$$T(\vec{e}_1, \vec{e}_1) = (\vec{e}_1, \vec{e}_2), T(\vec{e}_1, \vec{e}_2) = (\vec{e}_2, \vec{e}_2), T(\vec{e}_2, \vec{e}_1) = (\vec{e}_2, \vec{e}_2), \text{ and } T(\vec{e}_2, \vec{e}_2) = (\vec{e}_1, \vec{e}_1),$$

and let $M = C^{(2 \times 1)}$. *For* $r \in C$, *let* c_r *denote the corresponding scalar multiplication on* $C^{(2 \times 1)}$:

$$c_r : C^{(2 \times 1)} \to C^{(2 \times 1)}; \ \vec{x} \mapsto r\vec{x}.$$

Let $\mathbb{M} = (M; (c_r)_{r \in C}, +, T)$, *and let* $h : \mathbb{M} \to \mathbb{M}$ *be an endomorphism. Show that there is a matrix*

$$A = \begin{bmatrix} a_{11} & a_{12} \\ a_{21} & a_{22} \end{bmatrix}$$

such that for any $\vec{u} \in M$,

$$h(\vec{u}) = A\vec{u}.$$

Show that for this matrix A, we have

$$a_{11}^2 + a_{21}^2 = a_{11}$$

$$2a_{11}a_{21} = a_{21}$$

$$a_{21}^2 = a_{12}$$

$$a_{11}^2 + a_{21} = a_{22}$$

.

1.9 References

[BG] R. L. Bishop and S. I. Goldberg, *Tensor Analysis on Manifolds*, New York: MacMillan. 1968. (referenced on page 1)

[BeS] J. L. Bell and A. B. Slomson, *Models and Ultraproducts: An Introduction*, Amsterdam: North-Holland. 1969. (referenced on page 1)

[BS] S. Burris and H. P. Sankappanavar, *A Course in Universal Algebra*, New York: Springer-Verlag. 1981. (also now available online from the authors at http://www.thoralf.uwaterloo.ca/htdocs/ualg.html) (referenced on page 5)

[EFT] H. D. Ebbinghaus, J. Flum, and W. Thomas, *Mathematical Logic*, New York: Springer-Verlag. 1984. (referenced on page 1)

[E] H. B. Enderton, *A Mathematical Introduction to Logic*, 2nd. Ed. San Diego: Academic Press. 2002. (referenced on page

1)

[H] W. Hodges, *A Shorter Model Theory* Cambridge: Cambridge University Press. 2002. (further reading in advanced model theory)

[M] E. Mendelson, *Introduction to Mathematical Logic* Princeton: D. Van Nostrand. 1966. (further reading in introductory logic)

[Su] P. Suppes, *Axiomatic Set Theory* New York: Dover. 1972. (further reading in set theory)

Chapter 2

Languages and Satisfaction

2.1 Introduction

In proving theorems and solving problems, mathematicians and scientists use a language for communicating the proofs and solutions. Similarly, in computing, a modern computer uses a language to communicate with its programmers and users. Our multi-structures can be studied by mathematicians for their own sake, or they can be used to model computation systems, physical phenomena, etc. In order to make these uses of multi-structures, we shall make use of languages specifically designed for these

multi-structures. These languages are very similar to the
languages of classical mathematical logic.

2.2   Languages

A language $\mathcal{L}$ consists of the following:

1. A list, v_0, v_1, ..., v_n, ..., of denumerably many variables.

2. A list f_i, $i \in I$, of function symbols (also called
   operation symbols). Associated to each function
   symbol f_i is a pair (m_i, n_i) of nonnegative integers,
   called the arity of f_i.

3. A list, R_j, $j \in J$, of relation symbols (also called
   predicate symbols). Associated to each relation symbol
   R_j is a nonnegative integer α_j, called the arity of R_j.

4. The equality symbol, $=$.

5. The logical connectives, $\neg$ and $\wedge$ (read 'not' and 'and').

6. The punctuation symbols, ')' and '('. (Right and left
   parentheses, respectively.)

7. The (existential) quantifier symbol, $\exists$, which we read as 'there is'.

8. A list, π_i^k, $i \in I$, $k \in \{1, ..., n_i\}$, of projection symbols. For each $i \in I$ and $k \in \{1, ..., n_i\}$, the arity of π_i^k is n_i.

A language has terms, which we define as follows:

1. Any variable is a term.

2. For $i \in I$, if t_1, ..., t_{m_i} are terms, then $\pi_i^k f_i t_1 ... t_{m_i}$ is a term.

3. A string is a term iff it can be obtained by a finite number of applications of the preceeding two generation rules.

A language has atomic formulas, which we define by the following:

1. If t_1 and t_2 are terms, then the string $(t_1 = t_2)$ is an atomic formula.

2. If t_1, ..., t_{α_j} are terms, then $R_j t_1 ... t_{\alpha_j}$ is an atomic formula.

A language has formulas, which are given as follows:

1. Every atomic formula is a formula.

2. If φ and ψ are formulas and x is a variable, then

    (a) $(\neg\varphi)$, called the negation of φ, is a formula,

    (b) $(\varphi \wedge \psi)$ is a formula, which is called the conjunction of φ and ψ, and

    (c) $(\exists x)\varphi$ is a formula.

3. A string is a formula iff is can be obtained by finitely many applications of the preceeding rules.

Let x be a variable. Then the string $(\exists x)$ is called an x-quantifier. We say

1. The scope of the leading occurrence of the x-quantifier $(\exists x)$ in the formula $(\exists x)\varphi$ is $(\exists x)\varphi$.

2. The scope of any occurrence of the x-quantifier $(\exists x)$ in the formula $(\neg\varphi)$ is the same as its scope in the formula φ.

3. For any occurrence of the x-quantifier $(\exists x)$ in the formula φ, the scope of this occurrence in the formula $(\varphi \wedge \psi)$ is the same as its scope in φ.

4. For any occcurrence of the x-quantifier $(\exists x)$ in the formula ψ, the scope of this occurrence in the formula $(\varphi \wedge \psi)$ is the same as its scope in ψ.

5. For any occurrence of the x-quantifier $(\exists x)$ in the formula φ, the scope of this occurrence in the formula $(\exists y)\varphi$ is the same as its scope in the formula φ.

Note: When using parentheses, we will sometimes write '[' and ']', instead of '(' and ')', to improve readability for human readers. For example, instead of $((\varphi \wedge \psi) \wedge \xi)$, we might write $([\varphi \wedge \psi] \wedge \xi)$. However, in mathematical and computer applications, it is not strictly necessary to have two different "forms" of parentheses.

Example 2.2.1 *In the formula*

$$(\exists v_0) \left[\left(\pi_5^3 f_5 t_0 ... t_{n_5} = v_0 \right) \wedge (\exists v_1)(v_1 = v_2) \right],$$

the scope of the only occurrence of $(\exists v_1)$ *is* $(\exists v_1)(v_1 = v_2)$.

An occurrence of the variable x in the scope of a quantifier $(\exists x)$ is said to be a bound occurrence of the variable x. Occurrences of x that are not in the scope of any x-quantifier are free occurrences. If the variable x does not

occur in a formula φ, then any occurrence of an x-quantifier
in φ is vacuous. A formula in which every occurrence of
every variable is bound is a **sentence**.

Let φ be a formula, let x and y be variables, and suppose
that in φ, x does not occur in the scope of any occurrence
of a y-quantifier. Then the variable y is **free for** x **in** φ. If y
is free for x in φ and y does not occur free in φ, then
replacement of all free occurrences of the variable x in φ by
occurrences of y yields a formula φ_y^x which 'says' of y what
φ 'says' of x. More generally, given a term t and a variable
x, we define φ_t^x inductively, as follows:

1. First, we define u_t^x for a term u:

    (a) If u is x, then u_t^x is t.

    (b) If u is a variable $y \neq x$, then u_t^x is u (i.e. $u_t^x = y$).

    (c) If u is $\pi_i^k f_i t_1 ... t_{m_i}$, where t_1, ..., t_{m_i} are terms, then
    u_t^x is

    $$\pi_i^k f_i (t_1)_t^x ... (t_{m_i})_t^x.$$

2. Second, we define φ_t^x for an atomic formula φ:

    (a) If φ is $(t_1 = t_2)$, where t_1 and t_2 are terms, then φ_t^x

is the atomic formula

$$((t_1)_t^x = (t_2)_t^x).$$

(b) If φ is $R_j t_1 ... t_{\alpha_j}$, where t_1, ..., t_{α_j} are terms, then φ_t^x

is the atomic formula

$$R_j (t_1)_t^x ... (t_{\alpha_j})_t^x.$$

3. Finally, we define φ_t^x for non-atomic formulas φ as

follows:

(a) If φ is $(\neg \gamma)$, then φ_t^x is $(\neg \gamma_t^x)$.

(b) If φ is $(\gamma \wedge \psi)$, then φ_t^x is

$$(\gamma_t^x \wedge \psi_t^x).$$

(c) If φ is $(\exists x)\gamma$, then φ_t^x is $(\exists x)\gamma$.

(d) If φ is $(\exists y)\gamma$, where $y \neq x$, then φ_t^x is

$$(\exists y)\gamma_t^x.$$

Example 2.2.2 *Let φ be the formula*

$$(\exists y)[\neg(x = y)].$$

Then the formula φ_y^x is

$$(\exists y)[\neg(y = y)].$$

On the other hand, φ_x^x is φ.

Now we generalize the notion of a variable being "free for" another variable, to include terms, as follows:

1. If φ is atomic, then t is substitutable for x in φ.

2. If φ is $(\neg\psi)$, then t is substitutable for x in φ iff t is substitutable for x in ψ.

3. If φ is $(\psi \wedge \gamma)$, then t is substitutable for x in φ iff t is substitutable for x in both ψ and γ.

4. If φ is $(\exists y)\psi$, then t is substitutable for x in φ iff one of the following holds:

   (a) x does not occur free in φ, or

   (b) y does not occur in t and t is substitutable for x in ψ.

2.3   Satisfaction

Now, let $\mathbb{M}$ be a multi-structure of functional type $(m_i, n_i)_{i\in I}$ and of relational type $(\alpha_j)_{j\in J}$. Then the language $\mathcal{L}$ (which we have described in the preceeding section) and the multi-structure $\mathbb{M}$ are appropriate for one another.

(Loosely, this means that the sentences of the language $\mathcal{L}$ assert properties of the multi-structure $\mathbb{M}$.) Thus $\mathbb{M}$ is given by

$$\mathbb{M} = \left(M; \left(f_i^{\mathbb{M}} \right)_{i \in I}; \left(R_j^{\mathbb{M}} \right)_{j \in J} \right),$$

where for each $i \in I$, $f_i^{\mathbb{M}} : M^{m_i} \to M^{n_i}$, and for each $j \in J$, $R_j^{\mathbb{M}} \subseteq M^{\alpha_j}$, and f_i, $i \in I$, are the function symbols of $\mathcal{L}$, and R_j, $j \in J$, are the relation symbols of $\mathcal{L}$. Let $x : \omega \to M$. (That is, x is a sequence in the set M, indexed by the natural numbers $0, 1, 2, \ldots$) We will frequently denote such a sequence by

$$(x_k)_{k < \omega}, \text{ or } (x_0, x_1, \ldots).$$

We refer to x as a valuation in $\mathbb{M}$, and we can identify it with the function s defined on the set of variables by the rule

$$s(v_k) = x_k.$$

We extend s to the terms as follows: Let $\mathcal{T}$ be the set of terms of the language $\mathcal{L}$. Then $\check{s} : \mathcal{T} \to M$ is given by the following:

    1. For $k < \omega$, $\check{s}(v_k) = s(v_k) = x_k$.

2. **If $t_1, ..., t_{m_i}$ are terms, then**

$$\left(\check{s} \left(\pi_i^1 f_i t_1 ... t_{m_i} \right), ..., \check{s} \left(\pi_i^{n_i} f_i t_1 ... t_{m_i} \right) \right) = f_i^M \left(\check{s}(t_1), ..., \check{s}(t_{m_i}) \right).$$

Let x be a valuation in $\mathbb{M}$, let $k < \omega$, and let $a \in M$. Then we define $x(k/a) : \omega \to M$ by the rule

$$x(k/a)_l = \begin{cases} a & \text{if } l = k \\ x_l & \text{if } l \neq k \end{cases}.$$

Similarly, for the corresponding function s, if v is a variable and $a \in M$, then we define $s(v|a)$ by the rule

$$s(v|a)(w) = \begin{cases} a & \text{if } w = v \\ s(w) & \text{if } w \neq v \end{cases}.$$

Now we define the notations

$$\mathbb{M} \models_x \varphi, \text{ and } \mathbb{M} \models \varphi[s].$$

1. **For atomic formulas φ,**

   (a) **If φ is $(t_1 = t_2)$, then both $\mathbb{M} \models_x \varphi$, and $\mathbb{M} \models \varphi[s]$ mean that $\check{s}(t_1) = \check{s}(t_2)$ in M.**

   (b) **If φ is $R_j t_1 ... t_{\alpha_j}$, then both $\mathbb{M} \models_x \varphi$, and $\mathbb{M} \models \varphi[s]$ mean that $(\check{s}(t_1), ..., \check{s}(t_{\alpha_j})) \in R_j^M$.**

2. **For non-atomic formulas φ, ,**

(a) If φ is $(\neg\gamma)$, then both $\mathbb{M} \models_x \varphi$, and $\mathbb{M} \models \varphi[s]$ mean that it is not the case that $\mathbb{M} \models_x \gamma$.

(b) If φ is $(\gamma \wedge \psi)$, then both $\mathbb{M} \models_x \varphi$, and $\mathbb{M} \models \varphi[s]$ mean that both $\mathbb{M} \models_x \gamma$ and $\mathbb{M} \models_x \psi$.

(c) If φ is $(\exists v_k)\gamma$, then both $\mathbb{M} \models_x \varphi$, and $\mathbb{M} \models \varphi[s]$ mean that there is $a \in M$ such that

$$\mathbb{M} \models_{x(k/a)} \gamma.$$

Let us pause to introduce some abbreviations. For

$$\neg((\neg\gamma) \wedge (\neg\psi)),$$

we write

$$\gamma \vee \psi,$$

and for

$$(\neg\psi) \vee \psi,$$

we write

$$\gamma \rightarrow \psi.$$

Moreover, we abbreviate

$$(\gamma \rightarrow \psi) \wedge (\psi \rightarrow \gamma)$$

by

$$\gamma \leftrightarrow \psi.$$

Finally, for

$$\neg(\exists x)\neg\gamma,$$

we write

$$(\forall x)\gamma.$$

Our notation represents intuitive English as follows:

$$\neg\gamma \;\sim\; \text{``not } \gamma\text{''}$$

$$\gamma \wedge \psi \;\sim\; \text{``}\gamma \text{ and } \psi\text{''}$$

$$\gamma \vee \psi \;\sim\; \text{``}\gamma \text{ or } \psi\text{''} \text{ (in the inclusive sense)}$$

$$\gamma \rightarrow \psi \;\sim\; \text{``}\gamma \text{ implies } \psi\text{''}$$

$$\gamma \leftrightarrow \psi \;\sim\; \text{``}\gamma \text{ if and only if } \psi\text{''} \text{ (or ``}\gamma \text{ iff } \psi\text{'')}$$

$$(\exists x)\gamma \;\sim\; \text{``There is } x \text{ such that } \gamma\text{''}$$

$$(\forall x)\gamma \;\sim\; \text{``For all } x,\ \gamma\text{''}.$$

2.4   Truth, Sentences, and Universal Validity

If $\mathbb{M}$ is a multi-structure, φ is a formula, and x is a valuation in $\mathbb{M}$, and if

$$\mathbb{M} \models_x \varphi,$$

then we say that x satisfies φ in $\mathbb{M}$. Similarly, if s is defined by $s(v_k) = x_k$, then for $\mathbb{M} \models \varphi[s]$, we say that s satisfies φ in $\mathbb{M}$. In particular, if $\mathbb{M} \models_x \varphi$ for all valuations x in $\mathbb{M}$, then we say that φ is true in $\mathbb{M}$ (or that φ is valid in $\mathbb{M}$), and we write

$$\mathbb{M} \models \varphi.$$

(We also read the above as follows: "$\mathbb{M}$ satisfies φ".) It can be shown that if σ is a sentence, then $\mathbb{M} \models \sigma$ if and only if there is a valuation x in $\mathbb{M}$ such that $\mathbb{M} \models_x \sigma$. A formula φ such that $\mathbb{M} \models \varphi$ for every multi-structure $\mathbb{M}$, is universally valid.

2.5   Further Reading

Traditional developments of languages can be found in logic texts such as [BeS], [EFT], [E], [H] and [M]. In particular, [H] provides a more advanced treatment, but of traditional structures. The languages presented in universal algebra texts, such as [BS], for example, is an equational logic for traditional universal algebras. Set theoretic notation and axioms can be found, for example, in [S] or [Su].

2.6   Exercises

Exercise 2.6.1 *Let φ be the formula*

$$(\exists v_3)R_j v_1 v_3 v_{42} \rightarrow (\forall v_4)f_i v_1 v_3 v_4,$$

where the relation symbol R_j is ternary (i.e. it is 3-ary, which means that $\alpha_j = 3$) and the function symbol f_i is (3,1)-ary. In φ, which variables are

   1. in the scope of the v_3-quantifier?

   2. in the scope of the v_4-quantifier?

   3. free?

   4. free for v_{42}?

   5. free for v_4?

   6. in the scope of a v_1-quantifier?

Exercise 2.6.2 *Translate the following from English into a suitable formal language:*

"Fermat was a lawyer, not a mathematician."

Exercise 2.6.3 *Translate the following from its formal language into a suitable English sentence. (Note: R_j is a* binary *(i.e. 2-ary) relation symbol, f_{i_1} is a (2,2)-ary function symbol, and f_{i_2} is a binary (i.e. a (2,1)-ary) function symbol.)*

$$(\forall v_0)(\forall v_1)\left[R_j\pi^1_{i_1}f_{i_1}v_0v_1\pi^1_{i_2}f_{i_2}v_0v_1 \wedge R_j\pi^1_{i_2}f_{i_2}v_0v_1\pi^2_{i_1}f_{i_1}v_0v_1\right].$$

Exercise 2.6.4 *Give an example of a formula in which v_0 occurs both bound and free, and free for v_1.*

Exercise 2.6.5 *Give a* precise definition *of what it means for a symbol to be the kth symbol in a formula φ.*

Exercise 2.6.6 *Let R_j be a unary (i.e. 1-ary) relation symbol, and let f_i be a (2,3)-ary function symbol. For each of the following formulas φ, find a multi-structure $\mathbb{M}$ and a valuation x in $\mathbb{M}$ such that $\mathbb{M} \models_x \varphi$:*

1. *φ is $(\forall v_1)[\pi^1_i f_i v_0 v_1 = v_0]$,*

2. *φ is $(\forall v_1)[\pi^2_i f_i v_0 v_1 = v_0]$,*

3. *φ is $(\exists v_0)(\forall v_1)[\pi^1_i f_i v_0 v_1 = v_1]$,*

4. *φ is $(\exists v_0)(\forall v_1)[\pi^3_i f_i v_0 v_1 = v_1]$,*

5. *φ is $(\exists v_0)[R_j v_0 \wedge (\forall v_1)R_j\pi^1_i f_i v_0 v_1]$,*

6. *φ is $(\exists v_0)[R_j v_0 \wedge (\forall v_1)R_j\pi^3_i f_i v_0 v_1]$.*

Exercise 2.6.7 *Let φ be a formula that does not contain $\neg$, $\rightarrow$ or $\leftrightarrow$. Show that there are a multi-structure $\mathbb{M}$ and a valuation x in $\mathbb{M}$ such that $\mathbb{M} \models_x \varphi$.*

Exercise 2.6.8 *Let R_j be a unary relation symbol, and let f_i be a (2,3)-ary function symbol. For each of the following formulas φ, find a multi-structure $\mathbb{M}$ and a valuation x in $\mathbb{M}$ such that $\mathbb{M} \models_x (\neg \varphi)$:*

   1. *φ is* $(\forall v_1)\,[\pi_i^1 f_i v_0 v_1 = v_0]$,

   2. *φ is* $(\forall v_1)\,[\pi_i^2 f_i v_0 v_1 = v_0]$,

   3. *φ is* $(\exists v_0)(\forall v_1)\,[\pi_i^1 f_i v_0 v_1 = v_1]$,

   4. *φ is* $(\exists v_0)(\forall v_1)\,[\pi_i^3 f_i v_0 v_1 = v_1]$,

   5. *φ is* $(\exists v_0)\,[R_j v_0 \wedge (\forall v_1) R_j \pi_i^1 f_i v_0 v_1]$,

   6. *φ is* $(\exists v_0)\,[R_j v_0 \wedge (\forall v_1) R_j \pi_i^3 f_i v_0 v_1]$.

2.7 References

[BeS] J. L. Bell and A. B. Slomson, *Models and Ultraproducts: An Introduction*, **Amsterdam: North-Holland. 1969.** (general reference)

[EFT] H. D. Ebbinghaus, J. Flum, and W. Thomas, *Mathematical Logic*, **New York: Springer-Verlag. 1984.** (general reference)

[E] H. B. Enderton, *A Mathematical Introduction to Logic*, **2nd. Ed. San Diego: Academic Press. 2002.** (general reference)

[H] W. Hodges, *A Shorter Model Theory* **Cambridge: Cambridge University Press. 2002.** (further reading in advanced model theory)

[M] E. Mendelson, *Introduction to Mathematical Logic* **Princeton: D. Van Nostrand. 1966.** (further reading in introductory logic)

[Su] P. Suppes, *Axiomatic Set Theory* **New York: Dover. 1972.** (further reading in set theory)

[S] R. R. Stoll, *Sets, Logic and Axiomatic Theories*, San Francisco: W. H. Freeman and Company. 1961. (general reference)

Chapter 3

A Sound Axiom System

3.1 Introduction

We present here an axiom system for deriving consequences in our logic system. Some authors would call the system we will present a deductive calculus. We begin with a set of logical axioms. What we will actually present are axiom schema. Note that we will now typically leave out the outer parentheses. For example, in our first axiom scheme, we write $\varphi \to (\psi \to \varphi)$ instead of $[\varphi \to (\psi \to \varphi)]$. This is as harmless an abuse of notation as using abbreviations such as $\vee$ and $\to$.

3.2   The Axiom Schema

Our axiom schema are the following:

Axiom Scheme 1: For formulas φ and ψ,

$$\varphi \rightarrow (\psi \rightarrow \varphi)$$

   is an axiom.

Axiom Scheme 2: For formulas φ, ψ, and ξ,

$$[\varphi \rightarrow (\psi \rightarrow \xi)] \rightarrow [(\varphi \rightarrow \psi) \rightarrow (\varphi \rightarrow \xi)]$$

   is an axiom.

Axiom Scheme 3: For formulas φ and ψ,

$$[(\neg\varphi) \rightarrow (\neg\psi)] \rightarrow (\psi \rightarrow \varphi)$$

   is an axiom.

Axiom Scheme 4: For formulas φ and ψ,

$$(\varphi \wedge \psi) \rightarrow \varphi$$

   is an axiom.

Axiom Scheme 5: For formulas φ and ψ,

$$(\varphi \wedge \psi) \rightarrow \psi$$

   is an axiom.

Axiom Scheme 6: For formulas φ, ψ, and ξ,

$$(\xi \to \varphi) \to [(\xi \to \psi) \to (\xi \to [\varphi \wedge \psi])]$$

is an axiom.

Axiom Scheme 7: For formulas φ and ψ,

$$\varphi \to (\varphi \vee \psi)$$

is an axiom.

Axiom Scheme 8: For formulas φ and ψ,

$$\varphi \to (\varphi \vee \psi)$$

is an axiom.

Axiom Scheme 9: For formulas φ, ψ, and ξ,

$$(\varphi \to \xi) \to [(\psi \to \xi) \to ([\varphi \vee \psi] \to \xi)]$$

is an axiom.

Axiom Scheme 10: For any formula φ, any variable x, and any term t, if t is substitutable for x in φ, then

$$(\forall x)\varphi \to \varphi_t^x$$

is an axiom.

Axiom Scheme 11: For any formulas φ and ψ, and any variable x,

$$(\forall x)(\varphi \to \psi) \to [(\forall x)\varphi \to (\forall x)\psi]$$

is an axiom.

Axiom Scheme 12: For any formula φ, and any variable x, if x does not occur free in φ, then

$$\varphi \to (\forall x)\varphi$$

is an axiom.

Axiom Scheme 13: For any variable x,

$$x = x$$

is an axiom.

Axiom Scheme 14: For any formulas φ and ψ, and for any variables x and y, if φ is atomic and ψ is obtained from φ by replacing any number of occurrences of x by y, then

$$(x = y) \to (\varphi \to \psi)$$

is an axiom.

Axiom Scheme 15: For any axiom φ, any variable x,

$$(\forall x)\varphi$$

is an axiom.

It can be shown that each axiom which is generated by the preceeding schema is universally valid. This is a fundamental part of the proof of the Soundness Theorem, which, loosely speaking, guarantees that our logic system cannot derive falsehoods from truths.

3.3 Rules of Inference and Proofs

We have one fundamental rule of inference, which we call modus ponens:

For any formulas φ and ψ, ψ is an **immediate consequence** *of φ and $\varphi \rightarrow \psi$.*

We use modus ponens in the generation of proofs. Let Σ be a set of formulas and let φ be a formula. A proof (also called a deduction) of φ from Σ is a finite sequence φ_1, ..., φ_n of formulas such that

1. φ_n is φ (φ_n is the conclusion of the proof), and

2. For $k \leq n$, one of the following holds:

   (a) $\varphi_k \in \Sigma$,

   (b) φ_k is a logical axiom, or

   (c) there are $l, m < k$ such that φ_k is an immediate
       consequence of φ_l and φ_m.

In case there is a proof of φ from Σ, we say that φ is
provable from Σ, or φ is deducible from Σ, and we write

$$\Sigma \vdash \varphi.$$

Our rule modus ponens is postulated, and later we shall
describe how it contributes to the soundness of our
deduction system. For now, let us describe a derived rule of
inference. To do so, we first prove a theorem about
deductions.

Theorem 3.3.1 *Let Σ be a set of formulas, and let x be a
variable that does not occur free in any formula of Σ. Let
φ be a formula such that $\Sigma \vdash \varphi$. Then*

$$\Sigma \vdash (\forall x)\varphi.$$

Proof *We show this by induction. It is clear if φ is a logical axiom. Suppose that $\varphi \in \Sigma$. Then x does not occur in φ, so that*

$$\varphi \to (\forall x)\varphi$$

is a logical axiom. Thus $\Sigma \vdash \varphi$ and $\Sigma \vdash \varphi \to (\forall x)\varphi$. Using modus ponens, we get

$$\Sigma \vdash (\forall x)\varphi,$$

as desired. Now suppose that φ is obtained using modus ponens, say $\Sigma \vdash \psi$ and $\Sigma \vdash \psi \to \varphi$. Since the formula

$$(\forall x)(\psi \to \varphi) \to [(\forall x)\psi \to (\forall x)\varphi]$$

is a logical axiom, modus ponens yields

$$\Sigma \vdash (\forall x)\psi \to (\forall x)\varphi.$$

Then another application of modus ponens yields

$$\Sigma \vdash (\forall x)\varphi,$$

as desired.

For reasons that should now seem obvious, the preceeding theorem is called the Generalization Theorem, and it provides justification for using an additional (derived) rule of inference, known as generalization:

For any formula φ and any variable x, $(\forall x)\varphi$ is an immediate consequence of φ.

We note that the use of derived rules of inference shortens "proofs". Some lemmas prove useful in what follows, and their proofs are trivial, but we include them here for the novice reader.

Lemma 3.3.1 *Let Σ be a set of formulas, let φ be any formula, and let x and y be variables, such that y is free for x in φ. Then*

$$\Sigma \vdash (\forall x)\varphi \rightarrow \varphi_y^x.$$

Proof *Since y is free for x in φ, it is substitutable for x in φ, so the formula $(\forall x)\varphi \rightarrow \varphi_y^x$ is a logical axiom. The desired result follows.*

Lemma 3.3.2 *Let Σ be a set of formulas, let φ be a formula, and let x be a variable such that $\Sigma \vdash (\forall x)\varphi$. Then*

$$\Sigma \vdash \varphi.$$

Proof *Since x is substitutable for x in φ, and since φ_x^x is φ,*

$$\Sigma \vdash (\forall x)\varphi \rightarrow \varphi,$$

so by modus ponens,

$$\Sigma \vdash \varphi.$$

Now, some formulas are the "same as" others, except for the variables used. These are what we define next, inductively:

1. If φ is an atomic formula, then its only alphabetic variant is φ itself.

2. If φ is any formula, then the alphabetic variants of $(\neg\varphi)$ are the formulas of the form $(\neg\psi)$, where ψ is an alphabetic variant of φ.

3. If φ and ψ are formulas, then the alphabetic variants of $(\varphi \wedge \psi)$ are the formulas of the form $(\gamma \wedge \delta)$, where γ is an alphabetic variant of φ and δ is an alphabetic variant of ψ.

4. If φ is a formula and x and y are variables, and if y is free for x in φ, then the formula $(\exists y)\varphi_y^x$ is an alphabetic variant of the formula $(\exists x)\varphi$.

5. A formula ψ is an alphabetic variant of a formula φ if
   and only if finitely many applications of the preceeding
   rules show it to be so.

It is natural to expect that the property of "being an
alphabetic variant" is symmetric, but this turns out to be
more difficult to demonstrate than one might expect.

Proposition 3.3.1 *Let φ and ψ be formulas. If φ is an
alphabetic variant of ψ, then ψ is an alphabetic variant of
φ.*

Proof *This is trivial for atomic formulas, so suppose that
it is true for some set S of formulas, and let φ be $(\neg\gamma)$,
where $\gamma \in S$. Assume that φ is an alphabetic variant of ψ.
Since φ is $(\neg\gamma)$, ψ must be $(\neg\delta)$ for some formula δ. But
also, δ must be an alphabetic variant of γ. Since $\gamma \in S$, it
follows that γ is an alphabetic variant of δ, so the desired
result follows.*

*Now suppose that φ is $(\gamma \wedge \delta)$, where $\gamma, \delta \in S$, and let ψ be
such that φ is an alphabetic variant of ψ. Then ψ is
$(\mu \wedge \nu)$, for some formulas μ and ν. But then γ is an*

alphabetic variant of μ and δ is an alphabetic variant of ν,

so since $\gamma, \delta \in S$, it follows that μ is an alphabetic variant

of γ and ν is an alphabetic variant of δ, so that ψ is an

alphabetic variant of φ, as desired. Finally, suppose that x

is a variable and φ is $(\exists x)\gamma$, for some formula $\gamma \in S$.

Then, since φ is an alphabetic variant of ψ, φ is of the

form $(\exists y)\gamma_y^x$, where y is free for x in γ. But then (exercise)

$(\gamma_y^x)_x^y$ is γ and $(\exists x)(\gamma_y^x)_x^y$ is therefore ψ, so it follows that ψ

is an alphabetic variant of φ, as desired.

We can prove that *useful* alphabetic variants exist.

Theorem 3.3.2 *Let φ be a formula, let x be a variable, and*
let t be a term. Then there is an alphabetic variant ψ of φ,
such that

   1. $\{\varphi\} \vdash \psi$ and $\{\psi\} \vdash \varphi$, and

   2. t is substitutable for x in ψ.

Proof *If φ is atomic, then ψ must be φ, and it is clear that*
t is substitutable for x in ψ, in this case. If φ is $(\neg\gamma)$, then
we let ψ be $(\neg\delta)$, where δ is an alphabetic variant of γ such
that $\{\gamma\} \vdash \delta$, $\{\delta\} \vdash \gamma$, and t is substitutable for x in δ. Then

the desired result holds for ψ. If φ is $(\gamma \wedge \delta)$, then we let ψ be $(\beta \wedge \sigma)$, where β, σ are alphabetic variants of γ, δ, respectively, $\{\beta\} \vdash \gamma$, $\{\gamma\} \vdash \beta$, $\{\delta\} \vdash \sigma$, $\{\sigma\} \vdash \delta$, and t is substitutable for x in β and in σ. Then the desired result holds for ψ. Now, if φ is $(\forall y)\gamma$, then we distinguish two cases:

Case 1: *If either y does not occur in t or y is x, then we let ψ be $(\forall y)\delta$, where δ is an alphabetic variant of γ, $\{\delta\} \vdash \gamma$, $\{\gamma\} \vdash \delta$, and t is substitutable for x in δ. The desired result then holds for ψ.*

Case 2: *If y occurs in t and y is not x, then let δ be an alphabetic variant of γ such that $\{\delta\} \vdash \gamma$, $\{\gamma\} \vdash \delta$, and t is substitutable for x in δ. Let z be the first variable such that*

    1. z is not x,

    2. z does not occur in δ, and

    3. z does not occur in t.

Then we let ψ be

$$(\forall z)\delta_z^y.$$

It is then easy to see that t is substitutable for x in ψ.

We now show that $\{\varphi\} \vdash \psi$ and $\{\psi\} \vdash \varphi$. To see that

$\{\varphi\} \vdash \psi$, observe that by the inductive hypothesis,

$\{\gamma\} \vdash \delta$, so that $\{\varphi\} \vdash (\forall y)\delta$. Since z does not occur in

δ, we get

$$\{(\forall y)\delta\} \vdash \delta_z^y,$$

and then by generalization and modus ponens, it

follows that

$$\{\varphi\} \vdash (\forall z)\delta_z^y,$$

i.e. $\{\varphi\} \vdash \psi$, as desired.

Now, to see that $\{\psi\} \vdash \varphi$, note that $\{\psi\} \vdash (\delta_z^y)_y^z$, and

that $(\delta_z^y)_y^z$ is δ, whence $\{\psi\} \vdash \delta$. Also, by the induction

hypothesis, $\{\delta\} \vdash \gamma$, so that $\{\psi\} \vdash \gamma$. Then, since y

does not occur free in ψ, generalization yields

$\{\psi\} \vdash \varphi$, as desired.

Let φ be a formula. If φ has no quantifier, then it is open.
A closure of φ is

$$(\forall x_1)...(\forall x_n)\varphi,$$

where x_1, ..., x_n are the free variables in φ. We get the
following useful proposition.

Proposition 3.3.2 *Let φ be a formula, and let ψ be a closure of φ. Then for any set Σ of formulas, $\Sigma \vdash \varphi$ iff $\Sigma \vdash \psi$.*

Proof *It is enough to show that $\Sigma \vdash \varphi$ iff $\Sigma \vdash (\forall x)\varphi$, where x occurs free in φ. Thus suppose $\Sigma \vdash \varphi$ (and that x occurs free in φ). By generalization, $\Sigma \vdash (\forall x)\varphi$. Now, for the converse, suppose that $\Sigma \vdash (\forall x)\varphi$. We will show that $\Sigma \vdash \varphi$. To this end, let y be a variable that is free for x in φ. Since $(\forall x)\varphi \to \varphi_y^x$ is a logical axiom, it follows by modus ponens that $\Sigma \vdash \varphi_y^x$. Using generalization, we get*

$$\Sigma \vdash (\forall y)\varphi_y^x.$$

We may assume that x does not occur in φ_y^x. (We leave verification of this to the reader.) Thus $(\forall y)\varphi_y^x \to (\varphi_y^x)_x^y$ is a logical axiom, so that since $(\varphi_y^x)_x^y$ is φ, we get

$$\Sigma \vdash \varphi.$$

Now we will prove the Finiteness Theorem for our logic system. The proof is the same as the proof of the same theorem for classical first-order predicate calculus.

Theorem 3.3.3 *Let φ be a formula, and let Σ be a set of formulas. If $\Sigma \vdash \varphi$, then there is a finite set $\Sigma_0 \subseteq \Sigma$ such that $\Sigma_0 \vdash \varphi$.*

Proof *Let $(\varphi_1, ..., \varphi_n)$ be a proof of φ from Σ, and let $\Sigma_0 = \Sigma \cap \{\varphi_1, ..., \varphi_n\}$. Then Σ_0 is finite, and $(\varphi_1, ..., \varphi_n)$ is a proof of φ from Σ_0, so $\Sigma_0 \vdash \varphi$.*

The next result is sometimes called the Deduction Theorem.

Theorem 3.3.4 *Let Σ be a set of formulas, and let γ and φ be formulas. Then $\Sigma \cup \{\gamma\} \vdash \varphi$ iff $\Sigma \vdash \gamma \to \varphi$.*

Proof *We prove one direction, by induction, and leave the converse as an exercise. Thus, suppose that $\Sigma \cup \{\gamma\} \vdash \varphi$, and let $(\varphi_1, ..., \varphi_n)$ be a proof of φ from $\Sigma \cup \{\gamma\}$. We may assume that γ is not φ (Why?) and that φ is neither a logical axiom nor a member of Σ. Thus some φ_j is of the form $\varphi_k \to \varphi$ (i.e. φ is obtained as φ_n by the rule modus ponens). By the inductive hypothesis, $\Sigma \vdash \gamma \to \varphi_k$ and $\Sigma \vdash \gamma \to (\varphi_k \to \varphi)$, since φ_k and $\varphi_k \to \varphi$ occur earlier in the proof. It can be shown (exercise) that*

$$\{\gamma \to \varphi_k, \gamma \to (\varphi_k \to \varphi)\} \vdash \gamma \to \varphi,$$

so it follows that

$$\Sigma \vdash \gamma \rightarrow \varphi,$$

as claimed.

As a consequence, we get the Contraposition Rule:

Corollary 3.3.1 *Let Σ be a set of formulas, and let φ and ψ be formulas. Then $\Sigma \cup \{\varphi\} \vdash (\neg\psi)$ iff $\Sigma \cup \{\psi\} \vdash (\neg\varphi)$.*

Proof *We need only show that $\Sigma \cup \{\varphi\} \vdash (\neg\psi)$ implies $\Sigma \cup \{\psi\} \vdash (\neg\varphi)$. Thus suppose that $\Sigma \cup \{\varphi\} \vdash (\neg\psi)$. By the deduction theorem, we get*

$$\Sigma \vdash \varphi \rightarrow (\neg\psi).$$

It is easy to see (exercise) that

$$\{\varphi \rightarrow (\neg\psi)\} \vdash \psi \rightarrow (\neg\varphi),$$

so it follows that

$$\Sigma \vdash \psi \rightarrow (\neg\varphi).$$

But then by modus ponens,

$$\Sigma \cup \{\psi\} \vdash (\neg\varphi),$$

as claimed.

3.4 The Soundness Theorem

If φ is a formula such that $\emptyset \vdash \varphi$, then we say that φ is provable (from no assumptions), and we write

$$\vdash \varphi.$$

The following result is known as the Soundness Theorem.

Theorem 3.4.1 *Every provable formula is universally valid.*

Proof *We will outline the proof, and leave the details for the reader. The first task is to show that each logical axiom is universally valid. To do this, consider first the axioms that have no quantifiers, and then those in schema 10, 11 and 12, and then observe that if φ is universally valid, then $(\forall x)\varphi$ is also universally valid. The next task is to show that if φ and $\varphi \rightarrow \psi$ are both universally valid, then ψ is universally valid, so that modus ponens produces only universally valid immediate consequences when provided with universally valid data. This is enough to prove the desired result.*

As a consequence, a more general version of the Soundness Theorem follows.

Corollary 3.4.1 *Let Σ be a set of formulas, and let φ be a formula such that $\Sigma \vdash \varphi$. If $\mathbb{M}$ is a multi-structure, and if x is a valuation in $\mathbb{M}$ such that $\mathbb{M} \models_x \sigma$ for each $\sigma \in \Sigma$, then $\mathbb{M} \models_x \varphi$.*

Proof *Let $(\sigma_1, ..., \sigma_n, \varphi)$ be a proof of φ from Σ, and note that the formula*

$$(\sigma_1 \wedge ... \wedge \sigma_n) \rightarrow \varphi$$

is provable (exercise). Hence it is universally valid. Let $\mathbb{M}$ be a multi-structure, with a valuation x in $\mathbb{M}$, that satisfies each member of Σ. Then $\mathbb{M} \models_x \sigma_1 \wedge ... \wedge \sigma_n$, so that $\mathbb{M} \models_x \varphi$, as desired.

3.5   Proof by Contradiction; the Consistency Theorem

Let Σ be a set of formulas. If there is some formula φ such that $\Sigma \vdash \varphi$ and $\Sigma \vdash (\neg\varphi)$, then Σ is inconsistent. Otherwise, Σ is consistent. The following result forms the basis for proofs by contradiction (also referred to as proofs by reductio ad absurdum).

Corollary 3.5.1 *Let* Σ *be a set of formulas and let* φ *be a formula. If* $\Sigma \cup \{\varphi\}$ *is inconsistent, then* $\Sigma \vdash (\neg\varphi)$.

Proof *Assuming that* $\Sigma \cup \{\varphi\}$ *is inconsistent, let* γ *be a formula such that* $\Sigma \cup \{\varphi\} \vdash \gamma$ *and* $\Sigma \cup \{\varphi\} \vdash (\neg\gamma)$. *Then*

$$\Sigma \vdash \varphi \to \gamma \text{ and } \Sigma \vdash \varphi \to (\neg\gamma).$$

It is easy to see that

$$\{\varphi \to \gamma, \varphi \to (\neg\gamma)\} \vdash (\neg\varphi),$$

and the desried result follows.

Another way to state the Soundness Theorem is the following Consistency Theorem.

Theorem 3.5.1 *The empty set of formulas is consistent.*

Proof *Assume, to the contrary, that* $\emptyset$ *is not consistent, and let* φ *be a formula such that both* $\vdash \varphi$ *and* $\vdash (\neg\varphi)$. *Then both* φ *and* $(\neg\varphi)$ *are universally valid, so that for any multi-structure* $\mathbb{M}$, *we have both* $\mathbb{M} \models \varphi$ *and* $\mathbb{M} \models (\neg\varphi)$. *We leave it to the reader to explain why this is absurd.*

We now introduce some notation for more convenient statements of results. First, if Σ is a set of formulas and x is

a valuation in a multi-structure $\mathbb{M}$, then we say that the
pair $(\mathbb{M}, x)$ is a model of σ, or that $(\mathbb{M}, x)$ models Σ iff x
satisfies every formula $\sigma \in \Sigma$ in $\mathbb{M}$. We denote this
situation by

$$\mathbb{M} \models_x \Sigma.$$

If Σ is a set of sentences, then the reference to the valuation
x is unnecessary, so we drop it, and write merely

$$\mathbb{M} \models \Sigma.$$

Thus, for a formula φ,

$$\mathbb{M} \models \{\varphi\} \text{ iff } \mathbb{M} \models \varphi.$$

Now, for a set Σ of formulas, and for φ a formula, if every
multi-structure $\mathbb{M}$ that models Σ also satisfies φ, then we
write

$$\Sigma \models \varphi,$$

and we say that φ is a consequence of Σ, or that Σ entails φ.
Thus the more general version of the soundness theorem
can be stated simply, concisely, and symbolically as follows:

$$\Sigma \vdash \varphi \Rightarrow \Sigma \models \varphi.$$

In the case that $\emptyset \models \varphi$, we write

$$\models \varphi.$$

Thus $\models \varphi$ means that φ is universally valid.

3.6 Constants

One thing we have done differently from other authors is to not distinguish "constant symbols" in our language. Our reason for this is simple: We have already included the possibility of constants in the same way that it is typically done in universal algebra. To clarify, we will formally define the notion of a constant, and the reader will then see that the introduction of "constant symbols" would have been redundant. Specifically, in our setting, a constant symbol is, by definition, $\pi_i^k f_i$, where f_i is of arity $(0, n_i)$.

Thus, in a multi-structure $\mathbb{M}$, a constant symbol, c. is interpreted as a function

$$c^{\mathbb{M}} = \pi_{n_i, k} \circ f_i^{\mathbb{M}} : M^0 \rightarrow M,$$

and it "picks out" a distinguished member of M.

Once we have constants, or, more precisely, once we have identified our constants, we may use them in proofs. In

order to do so, we define replacement of a constant by a
variable as we did replacement of a variable by a term. We
leave the details of this definition to the reader. Suffice it
to say that we can in fact generalize this replacement
process to replace a term with a term, so that the notation
$\varphi_{t_1}^{t_2}$ makes sense. In particular, the next theorem makes use
of such notation in the case that t_1 is the variable y and t_2 is
the constant c.

Theorem 3.6.1 *Let Σ be a set of formulas, and let φ be a
formula. Let c be a constant symbol that does not occur in
Σ. Then there is a variable y that does not occur in φ,
such that*

$$\Sigma \vdash (\forall y)\varphi_y^c.$$

*Moreover, there is a proof of $(\forall y)\varphi$ from Σ in which c does
not occur.*

Proof *Let $(\varphi_1, ..., \varphi_n, \varphi)$ be a deduction of φ from Σ, and let
y be a variable that does not occur in this proof. We shall
demonstrate that the sequence*

$$\left((\varphi_1)_y^c, ..., (\varphi_n)_y^c, \varphi_y^c\right)$$

is a proof of φ_y^c from Σ. To do this, we need to show that

for each $k \leq n$, $\left((\varphi_1)_y^c, ..., (\varphi_k)_y^c \right)$ is a deduction from Σ. If

$\varphi_k \in \Sigma$ or φ_k is a logical axiom, this is trivial, so we leave

it as an exercise. Thus suppose that we have shown this

for all but the last "line" of the proof (so that our

demonstration is an induction argument), and suppose

that φ is obtained from φ_k and $\varphi_l = (\varphi_k \rightarrow \varphi)$ by modus

ponens. But this is almost trivial, for then $(\varphi_k \rightarrow \varphi)_y^c$ is

$(\varphi_k)_y^c \rightarrow \varphi_y^c$, so the induction hypothesis yields

$$\Sigma \vdash \varphi_y^c.$$

By the generalization theorem,

$$\Sigma \vdash (\forall y)\varphi_y^c,$$

as desired. Moreover, the proof we have exhibited does not

include the constant c, because we replaced each

occurrence, in the proof, of c by the variable y.

The preceeding result is sometimes called Generalization on
Constants. Some of its corollaries are especially useful. We
leave their proofs as exercises.

Corollary 3.6.1 *Let Σ be a set of formulas, and let φ be a*
formula. Let c be a constant that does not occur in either

Σ *or* φ. **If** $\Sigma \vdash \varphi_c^x$, *then* $\Sigma \vdash (\forall x)\varphi$, *and there is a deduction of* $(\forall x)\varphi$ *from* Σ *that avoids any occurrence of the constant* c.

Corollary 3.6.2 *Let* Σ *be a set of formulas, let* φ *and* ψ *be formulas, and let* c *be a constant that does not occur in* $\Sigma \cup \{\varphi, \psi\}$. **If**

$$\Sigma \cup \{\varphi_c^x\} \vdash \psi,$$

then there is a deduction of ψ *from* $\Sigma \cup \{(\exists x)\varphi\}$ *in which* c *does not occur.*

3.7 Exercises

Exercise 3.7.1 *Let Σ be a set of formulas, let φ be a formula, and let $\mathbb{M}$ be a multi-structure. If $\Sigma \vdash \varphi$, and if the closure of each formula in Σ is satisfied by $\mathbb{M}$, show that the closure of φ is satisfied by $\mathbb{M}$.*

Exercise 3.7.2 *Show that if R_j is a unary relation symbol and f_i is a binary function symbol, then*

$$\vdash (\exists v_0)(\exists v_1)\left[R_j \pi_i^1 f_i v_0 v_1 \rightarrow (\forall v_0)(\forall v_1) R_j \pi_i^1 f_i v_0 v_1\right].$$

(Note: A "binary function symbol" has arity (2,1).)

Exercise 3.7.3 *Show that any instance of Axiom Schema 1 through 9 that does not involve any quantifier occurrences is universally valid.*

Exercise 3.7.4 *Show that if x is a variable, φ is a formula with no x-quantifiers, t is a term, and t is substitutable for x in φ, then the axiom*

$$(\forall x)\varphi \rightarrow \varphi_t^x$$

(which is an instance of Axiom Schema 10) is universally valid.

Exercise 3.7.5 *Show that if φ and ψ are formulas with no x-quantifiers, then the axiom*

$$(\forall x)(\varphi \rightarrow \psi) \rightarrow [(\forall x)\varphi \rightarrow (\forall x)\psi]$$

is universally valid.

Exercise 3.7.6 *Show that if φ is a formula and x is a variable that does not occur free in φ, then the axiom*

$$\varphi \rightarrow (\forall x)\varphi$$

is universally valid.

Exercise 3.7.7 *Show that if x is a variable, then the formula*

$$x = x$$

is universally valid.

Exercise 3.7.8 *Show that any instance of Axiom Scheme 14 is universally valid.*

Exercise 3.7.9 *Show that if φ is universally valid, then so is $(\forall x)\varphi$. Explain why this shows that Axiom Scheme 15 produces only universally valid axioms.*

Exercise 3.7.10 *Show that modus ponens is* sound, *meaning that if φ and $\varphi \to \psi$ are universally valid, then ψ is universally valid.*

Exercise 3.7.11 *Let Σ be a set of formulas, and let φ be a formula. Let x be a variable that does not occur free in Σ. Show that if φ is a logical axiom and $\Sigma \vdash \varphi$, then $\Sigma \vdash (\forall x)\varphi$. (Hint: This is easy, because the hypothesis $\Sigma \vdash \varphi$ is redundant and superfluous.)*

Exercise 3.7.12 *Let the variable y be free for x in the formula γ. Show that $(\gamma_y^x)_x^y = \gamma$.*

Exercise 3.7.13 *Let t be a term, and let x be a variable. Use the Axiom of Choice to prove that there is a function, a, defined on the set $\mathcal{F}$ of formulas, such that*

1. If $\varphi \in \mathcal{F}$, then $a(\varphi)$ is an alphabetic variant of φ,

2. $\{a(\varphi)\} \vdash \varphi$ and $\{\varphi\} \vdash a(\varphi)$,

3. t is substitutable for x in $a(\varphi)$.

(Hint: Use induction as well as the Axiom of Choice.)

Exercise 3.7.14 *Verify that in the proof of the following result, we may assume that the variable x does not occur in the formula φ_y^x:*

Let φ be a formula, and let ψ be a closure of φ. Then for any set Σ of formulas, $\Sigma \vdash \varphi$ iff $\Sigma \vdash \psi$.

(Hint: *Use alphabetic variants.*)

Exercise 3.7.15 *Let the term t be substitutable for the variable x in the formula δ, and let z be the first variable such that*

 1. *z is not x,*

 2. *z does not occur in δ, and*

 3. *z does not occur in t.*

Let ψ be

$$(\forall z)\delta_z^y.$$

Show that t is substitutable for x in ψ.

Exercise 3.7.16 *Let Σ be a set of formulas, and let γ and φ be formulas. Show that if $\Sigma \vdash \gamma \to \varphi$, then*

$$\Sigma \cup \{\gamma\} \vdash \varphi.$$

Exercise 3.7.17 *Show that if φ and ψ are formulas, then*

$$\{\varphi \to (\neg\psi)\} \vdash \psi \to (\neg\varphi).$$

Exercise 3.7.18 *Prove the Soundness Theorem, in detail.*

Exercise 3.7.19 *Show that if $(\sigma_1, ..., \sigma_n, \varphi)$ is a proof of φ, then*

$$(\sigma_1 \wedge ... \wedge \sigma_n) \to \varphi$$

is provable.

Exercise 3.7.20 *Show that for any formulas φ and γ,*

$$\{\varphi \to \gamma, \varphi \to (\neg\gamma)\} \vdash (\neg\varphi).$$

Exercise 3.7.21 *Explain why it is not possible for a multi-structure $\mathbb{M}$ to satisfy both φ and $\neg\varphi$.*

Exercise 3.7.22 *Prove the following in detail:*

Let Σ be a set of formulas, and let φ be a formula. Let c be a constant symbol that does not occur in Σ. Then there is a variable y that does not occur in φ, such that

$$\Sigma \vdash (\forall y)\varphi_y^c.$$

Moreover, there is a proof of $(\forall y)\varphi$ from Σ in which c does not occur.

Exercise 3.7.23 *Prove the following:*

Let Σ be a set of formulas, and let φ be a formula. Let c be a constant that does not occur in either Σ or φ. If $\Sigma \vdash \varphi_c^x$, then $\Sigma \vdash (\forall x)\varphi$, and there is a deduction of $(\forall x)\varphi$ from Σ that avoids any occurrence of the constant c.

Exercise 3.7.24 *Prove the following:*

Let Σ be a set of formulas, let φ and ψ be formulas, and let c be a constant that does not occur in $\Sigma \cup \{\varphi, \psi\}$. If

$$\Sigma \cup \{\varphi_c^x\} \vdash \psi,$$

then there is a deduction of ψ from $\Sigma \cup \{(\exists x)\varphi\}$ in which c does not occur.

Exercise 3.7.25 *Prove that if Σ is a consistent set of formulas, then there is a set $\widehat{\Sigma}$ such that*

1. *$\Sigma \subseteq \widehat{\Sigma}$, and*

2. *for each formula φ, either $\varphi \in \widehat{\Sigma}$ or $(\neg\varphi) \in \widehat{\Sigma}$, but not both.*

Show that any such set $\widehat{\Sigma}$ is consistent.

Exercise 3.7.26 *Prove that*

$$\vdash [(\exists x)\varphi \lor (\exists x)\psi] \leftrightarrow (\exists x)(\varphi \lor \psi).$$

Exercise 3.7.27 *Prove that*

$$\vdash (\exists x)(\varphi \land \psi) \rightarrow [(\exists x)\varphi \land (\exists x)\psi].$$

Exercise 3.7.28 *Explain why the following formula is not always universally valid:*

$$[(\exists x)\varphi \land (\exists x)\psi] \rightarrow (\exists x)(\varphi \land \psi).$$

Exercise 3.7.29 *Assume that for any set Σ of formulas, if $\Sigma \models \varphi$, then $\Sigma \vdash \varphi$. Show, using this assumption, that if Σ is a consistent set of formulas, then it has a model.*

Exercise 3.7.30 *Assume that any consistent set of formulas has a model. Show, using this assumption, that if $\Sigma \models \varphi$, then $\Sigma \vdash \varphi$.*

Exercise 3.7.31 *Show that if φ, ψ and ξ are formulas, then*

$$\{\varphi \lor \psi\} \models \xi \Leftrightarrow \text{ both } \{\varphi\} \models \xi \text{ and } \{\psi\} \models \xi.$$

Exercise 3.7.32 *Show that*

$$\models (\forall x)(\varphi \land \psi) \leftrightarrow [(\forall x)\varphi \land (\forall x)\psi].$$

Exercise 3.7.33 *Give a formal definition of the notation*

$\varphi_{t_1}^{t_2}$, *where t_1 and t_2 are arbitrary terms.*

3.8 References

[BeS] J. L. Bell and A. B. Slomson, *Models and Ultraproducts: An Introduction*, Amsterdam: North-Holland. 1969. (general reference)

[EFT] H. D. Ebbinghaus, J. Flum, and W. Thomas, *Mathematical Logic*, New York: Springer-Verlag. 1984. (general reference)

[E] H. B. Enderton, *A Mathematical Introduction to Logic*, 2nd. Ed. San Diego: Academic Press. 2002. (general reference)

[H] W. Hodges, *A Shorter Model Theory* Cambridge: Cambridge University Press. 2002. (further reading in advanced model theory)

[M] E. Mendelson, *Introduction to Mathematical Logic* Princeton: D. Van Nostrand. 1966. (further reading in introductory logic)

[Su] P. Suppes, *Axiomatic Set Theory* New York: Dover. 1972. (further reading in set theory)

[S] R. R. Stoll, *Sets, Logic and Axiomatic Theories*, San Francisco: W. H. Freeman and Company. 1961. (general reference)

Chapter 4

Completeness

4.1 Introduction

We will show that our axiom system is complete. Roughly this means that all true statements are provable. Compare this to the Soundness Theorem, which roughly translates as "all provable statements are true". First, let us introduce another bit of terminology that is useful in stating one form of the Completeness Theorem. Specifically, we say that a set Σ of formulas is satisfiable if there is a multi-structure $\mathbb{M}$, and there is a valuation s in $\mathbb{M}$, such that for each $\varphi \in \Sigma$, $\mathbb{M} \models \varphi[s]$.

4.2   The Completeness Theorem and Its

Proof

We begin by proving the most general form of the
Completeness Theorem .

Theorem 4.2.1 *Let Σ be a set of formulas, and let φ be a
formula. Then the following hold:*

1. $\Sigma \models \varphi$ implies $\Sigma \vdash \varphi$, and

2. if Σ is consistent, then it is satisfiable.

Proof *We prove the second statement, and leave the first
as an exercise. Thus let Σ be a consistent set of formulas.
Add to the language $\mathcal{L}$ as many constants as there are
formulas of $\mathcal{L}$, say λ constants are added. Enumerate the
pairs*

$$(\varphi_\alpha, x_\alpha)_{\alpha < \lambda},$$

*where φ_α is a formula of the expanded language and x_α is
a variable. Observe that in the expanded language, which
we heretofore denote by $\widehat{\mathcal{L}}$, the set Σ is still a consistent
set (exercise). Let γ_0 be the formula*

$$[\neg(\forall x_0)\varphi_0] \to (\neg\varphi_0)_{c_0}^{x_0},$$

where c_0 is the first new constant that does not occur in

φ_0*. More generally, for $\alpha < \lambda$, let γ_α be*

$$[\neg(\forall x_\alpha)\varphi_\alpha] \rightarrow (\neg\varphi_\alpha)_{c_\alpha}^{x_\alpha},$$

where c_α is the first new constant that does not occur in

any member of the set

$$\{\gamma_\beta | \beta \leq \alpha\} \cup \{\varphi_{\alpha+1}\}.$$

Let

$$\widehat{\Sigma} = \Sigma \cup \{\gamma_\alpha | \alpha < \lambda\}.$$

Then (exercise) $\widehat{\Sigma}$ is consistent, so (another exercise, :-})

let Δ be a set of formulas such that

>   *1. Δ is consistent,*

>   *2. $\widehat{\Sigma} \subseteq \Delta$, and*

>   *3. for each formula φ of the language $\widehat{\mathcal{L}}$, either $\varphi \in \Delta$ or*
>   *$(\neg\varphi) \in \Delta$.*

Now let M be the set of terms of the language $\widehat{\mathcal{L}}$, and set

$$E^M = \left\{ (t_1, t_2) \in M^2 \,\middle|\, (t_1 = t_2) \in \Delta \right\}.$$

For each $j \in J$, set

$$R_j^M = \left\{ (t_1, ..., t_{\alpha_j}) \in M^{\alpha_j} \,\middle|\, R_j t_1 ... t_{\alpha_j} \in \Delta \right\}.$$

For each $i \in I$, define

$$f_i^{\mathrm{M}} : M^{m_i} \to M^{n_i}$$

by

$$f_i^{\mathrm{M}}(t_1, ..., t_{m_i}) = \left(\pi_i^1 f_i t_1...t_{m_i}, ..., \pi_i^{n_i} f_i t_1...t_{m_i} \right).$$

Finally, for each new constant symbol , c, let $c^{\mathrm{M}} = c$.

Now we define a valuation s in $\mathbb{M}$: s is the identity

function, i.e.

$$s(v_k) = v_k.$$

It can be shown that $\check{s}$ is the identity function on the set

of terms. For a formula φ of $\widehat{\mathcal{L}}$, let $\varphi_E^=$ be obtained by

replacing the equal sign with E. Then (exercise)

$$\mathbb{M} \models \varphi_E^=[s] \Leftrightarrow \varphi \in \Delta.$$

Now, from $\mathbb{M}$, we create a new multi-structure, $\widehat{\mathbb{N}} = \mathbb{M}/E^{\mathrm{M}}$,

whose elements are the equivalence classes of members of

M under the equivalence relation E^{M}. Since (exercise) E^{M}

is a congruence of $\mathbb{M}$, the natural map h is a

homomorphism from $\mathbb{M}$ onto $\widehat{\mathbb{N}}$. From this, it can be

shown that for each $\delta \in \Delta$, $\widehat{\mathbb{N}} \models \delta[h \circ s]$. Let $\mathbb{N}$ be the reduct

of $\widehat{\mathbb{N}}$ to the language $\mathcal{L}$. Then, since $\Sigma \subseteq \Delta$, it follows that *for each $\varphi \in \Sigma$,*

$$\mathbb{N} \models \varphi[h \circ s].$$

As a corollary, we get the more specific form of the

Completeness Theorem:

Corollary 4.2.1 *Every universally valid formula is*

provable.

4.3   Compactness

A very important consequence of the soundness,

completeness, and finiteness theorems is the Compactness

Theorem.

Theorem 4.3.1 *Let Σ be a set of formulas, and let φ be any*

formula.

  1. *If $\Sigma \models \varphi$, then there is a finite set $\Sigma_0 \subseteq \Sigma$ such that*

     $\Sigma_0 \models \varphi$.

  2. *If each finite subset of Σ is satisfiable, then Σ is*

     satisfiable.

Proof *We show the first of the two statements, and leave*

the other as an exercise. Thus suppose that $\Sigma \models \varphi$. By *completeness, $\Sigma \vdash \varphi$, and so by the Finiteness Theorem, let $\Sigma_0 \subseteq \Sigma$ be finite with $\Sigma_0 \vdash \varphi$. But then by the Soundness Theorem, $\Sigma_0 \models \varphi$. This is the desired result.*

4.4   Exercises

Exercise 4.4.1 *Prove the equivalence of the two parts of the Completeness Theorem.*

Exercise 4.4.2 *Prove the first part of the Completeness Theorem.*

Exercise 4.4.3 *Prove that the two parts of the Compactness Theorem are equivalent.*

Exercise 4.4.4 *Prove the second part of the Compactness Theorem.*

Exercise 4.4.5 *Use the Compactness Theorem to prove that a map can be coloured with four colours if and only if each of its finite submaps can be coloured with four colours.*

Exercise 4.4.6 *Let φ_2 be the formula*

$$\neg(v_0 = v_1).$$

Explain why the sentence

$$(\exists v_1)(\exists v_0)\varphi_2$$

is only satisfied by multi-structures that have at least two

elements.

Exercise 4.4.7 *Let φ_2 be the formula*

$$\neg(v_0 = v_1),$$

and let σ_2 be

$$(\exists v_1)(\exists v_0)\varphi_2.$$

For each $n < \omega$, let φ_{n+3} be

$$\varphi_{n+2} \wedge \left([\neg(v_0 = v_{n+2})] \wedge \ldots \wedge [\neg(v_{n+1} = v_{n+2})]\right),$$

and let σ_{n+3} be

$$(\exists v_{n+2})\ldots(\exists v_0)\varphi_{n+3}.$$

Explain why for each n, the sentence σ_{n+2} is satisfied only

in multi-structures having at least $n + 2$ elements.

Exercise 4.4.8 *Let Σ be a set of sentences such that for*

each $n < \omega$, there is a model $\mathbb{M}_n$ of Σ that has at least n

elements. Prove that Σ has a model $\mathbb{M}$ such that for each

$n < \omega$, $\mathbb{M}$ has at least n elements.

Exercise 4.4.9 *Show that a set Σ of sentences that has*

arbitrarily large finite models must have an infinite model.

Exercise 4.4.10 *Let Σ be a set of sentences that has an infinite model, of cardinality α, where the language $\mathcal{L}$ has at most α symbols. Show that for each infinite cardinal $\kappa > \alpha$, Σ has a model of cardinality κ.*

Exercise 4.4.11 *Prove that there is no set Σ of sentences such that $\mathbb{M} \models \Sigma$ iff $\mathbb{M}$ is a finite multi-structure.*

Exercise 4.4.12 *Prove that there is no set Σ of sentences such that $\mathbb{M} \models \Sigma$ iff $\mathbb{M}$ is a well-ordered multi-algebra. (A well-ordered set is a poset in which every nonempty subset has a least element. A well-ordered multi-algebra is a multi-structure $\mathbb{M}$ with only one fundamental relation, $\leq$, such that the reduct $(M, \leq)$ is a well-ordered set, and such that for each fundamental operation $f_i^{\mathbb{M}}$ of $\mathbb{M}$, if $a_1 \leq b_1$, ..., $a_{m_i} \leq b_{m_i}$ in $(M, \leq)$, then $k \in \{1, ..., n_i\}$ implies that $\pi_{n_i,k}(f_i^{\mathbb{M}}(a_1, ..., a_{m_i})) \leq \pi_{n_i,k}(f_i^{\mathbb{M}}(b_1, ..., b_{m_i})).)$*

Exercise 4.4.13 *Let Σ be a set of sentences, and let σ be a sentence such that if $\mathbb{M}$ is an infinite model of Σ, then $\mathbb{M} \models \sigma$. Show that there is a $k < \omega$ such that if $\mathbb{M}$ is a model of Σ with at least k elements, then $\mathbb{M} \models \sigma$. (Hint: If* *this were false, then the set $\Sigma \cup \{\neg\sigma\}$ would have arbitrarily*

large finite models. Explain why this cannot be the case.)

Exercise 4.4.14 *Show that there is no set Σ of sentences*

such that $\mathbb{M} \models \Sigma$ iff $\mathbb{M}$ is a finite group.

Exercise 4.4.15 *Let Σ be a set of sentences, in a countable*

language, which has a denumerable model. Prove that Σ

has models of any infinite cardinality. (Hint: Use the

Compactness Theorem, in an expanded language.)

Exercise 4.4.16 *Show that there is no set Σ of sentences*

such that $\mathbb{M} \models \Sigma$ iff $\mathbb{M}$ is a finite

(left,left)(3,2)-multi-group.

Exercise 4.4.17 *Assume that the language $\mathcal{L}$ has the*

following components:

   1. *the existential quantifier, $\exists$,*

   2. *for each relation A on the set of real numbers, a*
      corresponding predicate symbol , ρ_A, the arity of ρ_A
      being α_A, where $A \subseteq R^{\alpha_A}$, and

   3. *for each function $F : R^{m_F} \to R^{n_F}$, a function symbol*
      f_F.

Let $\rho_<$ denote ρ_A, where

$$A = \{(a, b) \in R^2 | a < b\},$$

and let $\mathbb{R}$ be the resulting multi-structure: Set

$$I = \{F : R^{m_F} \to R^{n_F} | m_F, n_F < \omega\},$$

and

$$J = \{A \subseteq R^{\alpha_A} | \alpha_A < \omega\},$$

and let

$$\mathbb{R} = \left(R; (f_i^{\mathbb{R}})_{i \in I}; (\rho_j^{\mathbb{R}})_{j \in J} \right),$$

where

$$f_i^{\mathbb{R}} = i : R^{m_i} \to R^{n_i}$$

and

$$\rho_j^{\mathbb{R}} = j \subseteq R^{\alpha_j}.$$

For each nullary operation

$$f_i^{\mathbb{R}} : R^0 \to R,$$

let $c_r = f_i$, where $r = f_i(\emptyset)$.

Now let Σ be the set of sentences given by

$$\Sigma = \{\sigma | \mathbb{R} \models \sigma\} \cup \{\rho_< c_r v_0 | r \in R\}.$$

Show that each finite subset of Σ is satisfiable. Conclude

that Σ is satisfiable. Let $\mathbb{M}$ be any model of Σ, and denote

by $<$ the relation $\rho_<^{\mathbb{M}}$, and for each $r \in R$, if $f_i = c_r$, then

denote $f_i^{\mathbb{M}}$ by r. (This is a harmless abuse of notation.)

Show that there is in $\mathbb{M}$ an element μ such that for each

$r \in R$,

$$r < \mu.$$

Explain why such an element μ can be properly referred to

as an "infinite" member of $\mathbb{M}$.

Exercise 4.4.18 *Let $\mathbb{R}$ and $\mathbb{M}$ be as above. Show that the*

operations $+^{\mathbb{M}}$, $-^{\mathbb{M}}$, $\cdot^{\mathbb{M}}$, and $\div^{\mathbb{M}}$ are field operations on M.

Exercise 4.4.19 *Let $\mathbb{R}$ and $\mathbb{M}$ be as above. Show that in $\mathbb{M}$*

there is an element $\varepsilon > 0$ such that for each $r \in R$ with

$r > 0$,

$$\varepsilon < r.$$

Explain why any such element ε can properly be referred

to as an "infinitesimal" element of $\mathbb{M}$.

4.5 References

[BeS] J. L. Bell and A. B. Slomson, *Models and Ultraproducts: An Introduction*, Amsterdam: North-Holland. 1969. (general reference)

[EFT] H. D. Ebbinghaus, J. Flum, and W. Thomas, *Mathematical Logic*, New York: Springer-Verlag. 1984. (general reference)

[E] H. B. Enderton, *A Mathematical Introduction to Logic*, 2nd. Ed. San Diego: Academic Press. 2002. (general reference)

[H] W. Hodges, *A Shorter Model Theory* Cambridge: Cambridge University Press. 2002. (further reading in advanced model theory)

[HL] A. E. Hurd and P. A. Loeb, *An Introduction to Nonstandard Real Analysis*, Orlando: Academic Press. 1985. (general reference for some exercises)

[M] E. Mendelson, *Introduction to Mathematical Logic* Princeton: D. Van Nostrand. 1966. (further reading in introductory logic)

[Su] P. Suppes, *Axiomatic Set Theory* New York: Dover. 1972. (further reading in set theory)

[S] R. R. Stoll, *Sets, Logic and Axiomatic Theories*, San Francisco: W. H. Freeman and Company. 1961. (general reference)

Chapter 5

Basic Model Theory for

Multi-Structures

5.1   Introduction

Here we introduce and briefly develop some of the basic
tools of model theory for multi-structures.

5.2   Extensions, Theories, and

Elementary Equivalence

Let $\mathbb{K}$ be a substructure of a multi-structure $\mathbb{M}$. Then we say that $\mathbb{M}$ is an extension of $\mathbb{K}$. Extensions play an important role in model theory, but before we delve into this, let us define some other useful notions.

Given a multi-structure $\mathbb{M}$, the theory of $\mathbb{M}$ is the following set of sentences:

$$Th(\mathbb{M}) = \{\sigma | \mathbb{M} \models \sigma\}.$$

If two multi-structures, $\mathbb{M}_1$ and $\mathbb{M}_2$, have the same theory, then we say they are elementarily equivalent, and we write

$$\mathbb{M}_1 \equiv \mathbb{M}_2.$$

It is easy to see (exercise) that isomorphic multi-structures are elementarily equivalent, but the converse fails.

Theorem 5.2.1 *Let $\mathbb{M}_1$ be an infinite multi-structure of cardinality λ, and let $\mu > \lambda$. Then $\mathbb{M}_1$ has an elementarily equivalent extension $\mathbb{M}_2$ of cardinality μ, which is therefore not isomorphic to $\mathbb{M}_1$.*

Proof *For each $\alpha < \mu$, let c_α be a constant symbol not in the original language $\mathcal{L}$, and let $\widehat{\mathcal{L}}$ be the language obtained from $\mathcal{L}$ by adding the constants c_α, $\alpha < \mu$. Let*

$$\Sigma = Th(\mathbb{M}_1) \cup \{\neg(c_\alpha = c_\beta) | \alpha < \beta < \mu\}.$$

Then for any finite subset Σ_0 of Σ, there is a valuation x in $\mathbb{M}_1$ such that $\mathbb{M}_1 \models_x \Sigma_0$, so Σ has a model, $\widehat{\mathbb{M}_2}$. Let $\mathbb{M}_2$ be the reduct of $\widehat{\mathbb{M}_2}$ to the language $\mathcal{L}$. This model $\mathbb{M}_2$ is the desired extension of $\mathbb{M}_1$ that is elementarily equivalent to $\mathbb{M}_1$. (Exercise)

The proof of the following useful proposition is left as an exercise for the reader.

Proposition 5.2.1 *Let $\mathbb{M}_1$ be a substructure of $\mathbb{M}_2$, let φ be an atomic formula of $\mathcal{L}$, and let $x \in M_1^\omega$ be a valuation in $\mathbb{M}_1$. Then*

$$\mathbb{M}_1 \models_x \varphi \text{ iff } \mathbb{M}_2 \models_x \varphi.$$

Let $\mathbb{K}$ be a substructure of a multi-structure $\mathbb{M}$. Then $\mathbb{K}$ is an elementary substructure of $\mathbb{M}$ provided that for each formula φ of $\mathcal{L}$ and for each valuation x in $\mathbb{K}$,

$$\mathbb{K} \models_x \varphi \text{ iff } \mathbb{M} \models_x \varphi.$$

In this case, we also say that $\mathbb{M}$ is an elementary extension of $\mathbb{K}$. An embedding $h : \mathbb{M}_1 \rightarrowtail \mathbb{M}_2$ is an elementary embedding iff for each formula φ of $\mathcal{L}$ and for each valuation x in $\mathbb{M}_1$,

$$\mathbb{M}_1 \models_x \varphi \text{ iff } \mathbb{M}_2 \models_{hox} \varphi.$$

In case there exists an elementary embedding from $\mathbb{M}_1$ into $\mathbb{M}_2$, we say $\mathbb{M}_1$ is elementarily embeddable into $\mathbb{M}_2$.

Here is another nice result; we again leave its proof to the reader:

Proposition 5.2.2 *If* $h : \mathbb{M}_1 \to \mathbb{M}_2$ *is an isomorphism, then* h *is an elementary embedding.*

5.3   Chains and Elementary Chains

A chain of multi-structures is an indexed collection

$$(\mathbb{M}_\xi)_{\xi < \lambda},$$

where λ is a cardinal number and for any $\xi_1 \leq \xi_2 < \lambda$, $\mathbb{M}_{\xi_1}$ is a substructure of $\mathbb{M}_{\xi_2}$. Such a chain is an elementary chain if for $\xi_1 \leq \xi_2 < \lambda$, $\mathbb{M}_{\xi_2}$ is an elementary extension of $\mathbb{M}_{\xi_1}$.

A major result about elementary chains is the following:

Theorem 5.3.1 *Let* $(\mathbb{M}_\xi)_{\xi<\lambda}$ *be an elementary chain, and let* $\mathbb{M}$ *be defined as follows:*

1. $M = \bigcup_{\xi<\lambda} M_\xi,$

2. *for* $i \in I$, $f_i^{\mathbb{M}} = \bigcup_{\xi<\lambda} f_i^{\mathbb{M}_\xi}$, *and*

3. *for* $j \in J$, $R_j^{\mathbb{M}} = \bigcup_{\xi<\lambda} R_j^{\mathbb{M}_\xi}$,

and

$$\mathbb{M} = \left(M; \left(f_i^{\mathbb{M}} \right)_{i \in I}; \left(R_j^{\mathbb{M}} \right)_{j \in J} \right).$$

Then for each $\xi < \lambda$, $\mathbb{M}$ *is an elementary extension of* $\mathbb{M}_\xi$.

Proof *Let* S *be the set of all formulas* φ *such that if* $\xi < \lambda$, *and* $x : \omega \to M_\xi$ *is any valuation, then*

$$\mathbb{M}_\xi \models_x \varphi \text{ iff } \mathbb{M} \models_x \varphi.$$

Then, from a previous proposition, it follows that S *contains all atomic formulas. Suppose* $\varphi, \psi \in S$. *Then for any valuation* $x : \omega \to M_\xi$, *we have*

$$\mathbb{M} \models_x (\neg\varphi) \text{ iff } \mathbb{M}_\xi \not\models_x \varphi \text{ iff } \mathbb{M} \not\models_x \varphi \text{ iff } \mathbb{M} \models_x (\neg\varphi),$$

and

$$\mathbb{M}_\xi \models_x (\varphi \wedge \psi) \text{ iff } (\mathbb{M}_\xi \models_x \varphi \text{ and } \mathbb{M}_\xi \models_x \psi)$$

iff $(\mathbb{M} \models_x \varphi$ **and** $\mathbb{M} \models_x \psi)$ **iff** $\mathbb{M} \models_x (\varphi \wedge \psi)$.

Let v_k be any variable. If $\mathbb{M}_\xi \models_x (\exists v_k)\varphi$, then for some $a \in M_\xi$,

$$\mathbb{M}_\xi \models_{x(k/a)} \varphi,$$

so that

$$\mathbb{M} \models_{x(k/a)} \varphi,$$

whence

$$\mathbb{M} \models_x (\exists v_k)\varphi.$$

We leave the converse implication as an exercise for the reader.

5.4   Exercises

Exercise 5.4.1 *Prove that isomorphic multi-structures are elementarily equivalent.*

Exercise 5.4.2 *Prove the following statement:*

Let $\mathbb{M}_1$ be a substructure of $\mathbb{M}_2$, let φ be an atomic formula of $\mathcal{L}$, and let $x \in M_1^\omega$ be a valuation in $\mathbb{M}_1$. Then

$$\mathbb{M}_1 \models_x \varphi \ \textit{iff} \ \mathbb{M}_2 \models_x \varphi$$

.

Exercise 5.4.3 *Prove the following statement:*

If $h : \mathbb{M}_1 \to \mathbb{M}_2$ is an isomorphism, then h is an elementary embedding.

Exercise 5.4.4 *Let $(\mathbb{M}_\xi)_{\xi<\lambda}$ be an elementary chain, and let $\mathbb{M}$ be defined as follows:*

1. $M = \bigcup_{\xi<\lambda} M_\xi$,

2. for $i \in I$, $f_i^{\mathbb{M}} = \bigcup_{\xi<\lambda} f_i^{\mathbb{M}_\xi}$, and

3. for $j \in J$, $R_j^{\mathbb{M}} = \bigcup_{\xi<\lambda} R_j^{\mathbb{M}_\xi}$, *and*

$$\mathbb{M} = \left(M; \left(f_i^{\mathbb{M}} \right)_{i \in I}; \left(R_j^{\mathbb{M}} \right)_{j \in J} \right).$$

Let $\xi < \lambda$, and let x be a valuation in $\mathbb{M}_\xi$, and let v_k be any variable. Prove that if $\mathbb{M} \models_x (\exists v_k)\varphi$, then $\mathbb{M}_\xi \models_x (\exists v_k)\varphi$.

Exercise 5.4.5 *Find multi-structures $\mathbb{M}_1$ and $\mathbb{M}_2$ such that $\mathbb{M}_1 \equiv \mathbb{M}_2$, but $\mathbb{M}_1$ is not elementarily embeddable in $\mathbb{M}_2$ and $\mathbb{M}_2$ is not elementarily embeddable in $\mathbb{M}_1$.*

Exercise 5.4.6 *Show that the real numbers, as an ordered set, form an elementary extension of the rationals.*

Exercise 5.4.7 *Prove that the relation $\equiv$ (elementary equivalence) is reflexive, symmetric, and transitive. (i.e. It is an equivalence relation.)*

Exercise 5.4.8 *Prove that if $\mathbb{M}_1$ and $\mathbb{M}_2$ are multi-structures with $Th(\mathbb{M}_1) \subseteq Th(\mathbb{M}_2)$, then $Th(\mathbb{M}_2) = Th(\mathbb{M}_1)$, i.e. $\mathbb{M}_1 \equiv \mathbb{M}_2$.*

5.5   References

[BeS] J. L. Bell and A. B. Slomson, *Models and Ultraproducts: An Introduction*, **Amsterdam: North-Holland. 1969.** (general reference)

[EFT] H. D. Ebbinghaus, J. Flum, and W. Thomas, *Mathematical Logic*, **New York: Springer-Verlag. 1984.** (general reference)

[E] H. B. Enderton, *A Mathematical Introduction to Logic*, **2nd. Ed. San Diego: Academic Press. 2002.** (general reference)

[H] W. Hodges, *A Shorter Model Theory* **Cambridge: Cambridge University Press. 2002.** (further reading in advanced model theory)

[HL] A. E. Hurd and P. A. Loeb, *An Introduction to Nonstandard Real Analysis*, **Orlando: Academic Press. 1985.** (general reference for some exercises)

[M] E. Mendelson, *Introduction to Mathematical Logic* **Princeton: D. Van Nostrand. 1966.** (further reading in introductory logic)

[Su] **P. Suppes,** *Axiomatic Set Theory* **New York: Dover. 1972.** (further reading in set theory)

[S] **R. R. Stoll,** *Sets, Logic and Axiomatic Theories*, **San Francisco: W. H. Freeman and Company. 1961.** (general reference)

Index

.

www.ingramcontent.com/pod-product-compliance
Lightning Source LLC
Chambersburg PA
CBHW020357100426
42812CB00001B/91